Intersections of Religion, Education, and a Sustainable World

Sally Windsor • Olof Franck
Editors

Intersections of Religion, Education, and a Sustainable World

Springer

Editors
Sally Windsor
Department of Pedagogical, Curricular and Professional Studies, Faculty of Education
The University of Gothenburg
Gothenburg, Sweden

Olof Franck
Department of Pedagogical, Curricular and Professional Studies, Faculty of Education
The University of Gothenburg
Gothenburg, Sweden

ISBN 978-3-031-81811-0 ISBN 978-3-031-81809-7 (eBook)
https://doi.org/10.1007/978-3-031-81809-7

This work was supported by The University of Gothenburg, Mälardalen University, Malmö University, The University of Helsinki and The University of Oslo.

This Springer imprint is published by the registered company Springer Nature Switzerland AG
The registered company address is: Gewerbestrasse 11, 6330 Cham, Switzerland

Foreword

Sustainability challenges link our innermost hopes and fears, with worlds inhabited by untold known and unknown relationships. While these relationships at times have been depicted as stable, the past, present, and future are now seemingly in a state of flux. The good intentions of the past might have brought some groups prosperity, but the consequences can be fatal for environment and humans in both the present and future. And while the modernist future had promised us happiness and never-ending progress, this narrative now seems harder than ever to believe in.

Sustainability challenges and especially the changes that are happening to the climate are both globally dispersed in the most abstract manner and painfully present in the most mundane everyday practices. This necessitates a fluid approach to sustainability that does not seek to limit our efforts to understand and act to specific disciplines. Engaging with sustainability challenges as individuals and societies is thus much more than technical fixes or questions of behaviour modification. It is existential, as it challenges our beliefs and practices at the most fundamental level. Drawing on Kundera's notion of kitsch, it is no longer possible to insist on singular (modernistic or otherwise) solutions to the human predicament, in what has been coined the Anthropocene: Humans have entered the global scene, shaping our planet and at the same time losing control of a world that is seemingly spinning out of control. It must now be considered kitsch to hide in a corner of our educational disciplines in order to understand and engage the challenges that we face. But working across disciplines is not easy, and demands profound levels of insights into core bodies of disciplinary, practice, and research-knowledge and the will to go beyond those.

This book thoughtfully engages with an obvious and underdeveloped relationship between sustainability education and religious education. Throughout the chapters, the book engages with specific religious traditions, their traits, and how they can shed light on both great potentials for understanding how a sustainable development could be viable and how it is possible to understand the range of barriers that naturally forms when we have to change faster than culture and practices develop. And thus, the book moves beyond the role of specific religious traditions and poses the question of how we can understand sustainability as a religious question. A question relating to how we understand ourselves, as individuals and collectives, to be part of the world and how the world is a part of us and enacts with both our values, norms, habits, practices, hopes, and despair.

As a researcher coming from the field of environmental and sustainability education, I find that this book offers a strong foray into a much needed transdisciplinary take on education that transcends disciplines in the best of ways. That being, without subjugating the strong disciplinary bodies of knowledge to an overarching position or vice versa. Within education and at a broader societal level, there are seemingly no end to calls for a more deep-rooted and existential engagement with the sustainability challenges that manifest around and within us. This book helps by showcasing how we as human beings have always engaged with a wondrous, changing, and at times threatening world and how a meeting of sustainability and religion in education helps us navigate past, present, and future challenges.

Jonas Andreasen Lysgaard
Associate Professor in Environmental
and Sustainability Education Research
Danish School of Education
Aarhus University, Copenhagen, Denmark
jonas@edu.au.dk

Contents

Chapter 1
The Fertile Ground Between Sustainability and Religious Education

Sally Windsor and Olof Franck

1.1 Introduction

Research on, and discussion of, how sustainability is considered in Religious Education (RE) of various traditions is somewhat rare. Similarly, religion and RE is little acknowledged within current conceptions of Education for Sustainable Development (ESD[1]). This may seem strange in current times given that questions of environmental, social, cultural, and economic sustainability (both historically and in the present) play such a prominent role in theology, creed, ethics, and religious practices. The purpose of education has been described by Kemmis (2023), as "helping people to live well in a world worth living in" (p. 13). Living well in a world that allows all beings – human and non-human – to thrive (Kaukko et al., 2023), is not a new idea, but one that is central to education throughout time and to both RE and sustainability education in all their forms.

[1] The terms sustainability education and ESD are used interchangeably in this chapter. We use ESD as it is a globally recognised term for sustainability education that "gives learners of all ages the knowledge, skills, values and agency to address interconnected global challenges including climate change, loss of biodiversity, unsustainable use of resources, and inequality" (UNESCO, 2022). However, we wish to recognise the critiques of ESD as: a form of western developmental optimism (Hellberg & Knutsson, 2018); and an impossible ideal (see for example, Jickling, 1992; Pedersen et al., 2022), that has moved towards instrumentalism (see for example Lotz-Sisitka, 2004) in which there has been a narrowing of the view of what counts as education for sustainability.

S. Windsor (✉) · O. Franck
Department of Pedagogical, Curricular and Professional Studies, Faculty of Education, The University of Gothenburg, Gothenburg, Sweden
e-mail: sally.windsor@gu.se

S. Windsor, O. Franck (eds.), *Intersections of Religion, Education, and a Sustainable World*, https://doi.org/10.1007/978-3-031-81809-7_1

The right to practice religion, and by extension participate in religious education, is protected by a number of international treaties. For example, Article 18 in the *International Covenant on Civil and Political Rights* (UN, 1996) states that freedom of religion is a fundamental human right, where people are free to learn, practice, and teach a religion of their choice in both public and in private. The *Religion Declaration* (Article 4) (General Assembly resolution 36/55, 1981) and the *Declaration on the Rights of Persons Belonging to National or Ethnic, Religious and Linguistic Minorities* (UNHRC, 1992) require that signatories and member states protect the cultures and identities of minorities as this contributes to social and political stability. McDougall (2016/2010) shows us that the ongoing disregard of minority rights is where social injustice begins and leads to social instability and the outbreak of violence. Therefore, protecting minority rights, cultures, and identities through RE is critical in aiding security and social cohesion and preventing conflict (Khaled, 2024). RE may offer pathways to a more socially cohesive society. At the same time, sustainability education is increasingly focused on these aspects of living on planet earth, particularly in relation to inequality. ESD and RE can offer critical frameworks or approaches where various dogmatic, paternalistic and dividing roles of religion and capitalist society are brought in focus (see for example Franck, 2017a).

We argue that an education that is focused on ethics and values "has a natural place within both sustainability education and RE" (Franck, 2017b, p. 2). Many, even most, RE curricula include ethical and values related content, and religious scholars and clerics generally take their role as ethical teachers seriously. In fact, ethical and value-related themes with clear relevance to sustainability themes in religious and secular life views and traditions, are included as core content in many RE curricula around the world (see for example Rothgangel et al., 2014). In 1975 Grimmitt and Read distinguished different types of learning in RE: learning *about* religion, learning *from* religion and learning *in* religion. We see these types of learning in the debate around sustainability education too. Sustainability education should focus on learning *about* sustainability issues, as well as learning *from* and *by* living sustainably. This distinction is also useful in thinking about the integration of sustainability and religious education in terms of enhancing ethical literacy.

Although the term 'ethical literacy' is not used frequently (Osbeck, 2017), it is a term that we believe sits at the purpose-nexus of RE and sustainability education. To strive for ethical literacy is to strive for the ability to understand ethical concepts, to interpret and understand how such concepts work in ethical argumentation, and to develop a critical analysis of arguments for and against different ethical positions. In other words, when people are ethically literate they have developed the ability to communicate, interact and take action in ethical ways (Franck & Osbeck, 2017). Ethical literacy, when conceived in this way is analogous to Nussbaum's (2008) concept of the 'sympathetic imagination'. Nussbaum argues that it is necessary to

> Cultivate in ourselves a capacity for sympathetic imagination that will enable us to comprehend the motives and choices of people different from ourselves, seeing them not as forbiddingly alien and other, but as sharing many problems and possibilities with us. Differences

> of religion, gender, race, class, and national origin make the task of understanding harder, since these differences shape not only the practical choices people face but also their "insides": their desires, thoughts, and ways of looking at the world (Nussbaum, 2008, p. 144)

This book on religion, education and sustainability is intended to contribute to the debates on how sustainability issues can be integrated into RE. Teaching about religion, philosophies of life and ethics takes different forms in different places. These different forms of RE demonstrate the varied possibilities to integrate the multidimensional concept of sustainable development. The contributions of this book show, in various ways, how such multidimensionality can take concrete forms. As such, there are four main aims of this volume:

1. To enhance the conceptualisation of how education in various religious traditions can advance theoretical and conceptual understandings of sustainability and sustainable development. There is often a lack of clarity, or explicitness on sustainability in RE, and vice versa, a lack of understanding of the role religion can play in sustainability education. With this cloudiness educators may miss opportunities to look more closely at what teachings different religions can, and have historically offered, when it comes to inspiring a theoretical, existential understanding of the concept of sustainable development. The interpretations of what such understandings can mean in terms ethical consequences and of moral practices are therefore, very often under-scrutinised.
2. To gather research on teacher education and pedagogy for incorporating sustainability education into different forms of religious education. The relative silence on sustainability in RE implies that teachers who teach religion and ethics are not provided the support they need to develop teaching that integrates environmental, social, cultural and economic sustainability issues. We believe more research is needed on how pedagogical methods and strategies for including and updating ESD issues in RE classrooms can be developed.
3. To build a global multi-faith community of those interested in the intersection of sustainability and religious education. As young people navigate an increasingly precarious world more research is needed on how education can provide space for critical approaches and a belief in change, as well as a sense of hope that can carry towards the future; an argument that has been maintained for decades (see for example Dasberg, 1983). It is therefore critically important that questions of sustainability are considered in all forms of religious education which is why we sought contributions from all over the world and from many religious traditions. Additionally, religion's contribution to the general discussion on how education should integrate global agreements for the future such as the Agenda 2030 and the 17 UN sustainable development goals is needed.
4. To deepen understandings of the nexus between various religious education traditions and sustainability and contribute to global teaching about ethics, values and philosophical thoughts on life. We anticipate that in the long term such research can influence analysis and discussion of how policy documents for the teaching of, in and about religions, philosophy and ethics are designed in relation to issues of sustainable development.

The chapters in this book are varied in terms of country of the authors, the focus and the religious perspectives they emerge from, yet there are a number of common themes. The chapters include an articulation of conceptions of sustainability historically and in contemporary times within the various religious traditions, and share a focus on social justice, and crucially for this handbook, a focus on questions of living an ethical life: religious ethics, relational ethics, environmental ethics. There are chapters that aim to develop pedagogies for the integrated teaching of religion and sustainability, describing research conducted in classrooms that draws upon specific resources, both religious and secular to demonstrate how this can be done. Finally, there are chapters on Indigenous spirituality and religions, creation narratives and, importantly RE's relationship to the natural world.

1.2 The Chapters That Follow

In Chap. 2, **Ole Andreas Kvamme** demonstrates how religious education must be reconsidered, given the current sustainability challenges involving the climate crisis, ecological degradation and global injustice, that prompt a rethinking of pedagogy and education. Kvamme approaches this question by critically analysing three well-known and influential accounts of religious education, developed by Andrew Wright, Robert Jackson and Wanda Alberts. By paying attention to their strategies that make visible and thus justify the purpose of such teaching, Kvamme finds that all three provide guidance on how to accommodate the ecological, existential and political contexts in which religious education takes place. In this way, he argues, that each have relevant contributions to make to a rethinking of religious education to better speak to sustainability in our time.

In the third chapter, **Mohammad Fazlhashemi** conducts a critical examination of Muslim ecotheology, drawing on the form and expression of the Iranian philosopher Sayyid Hossein Nasr's thoughts. Fazlhashemi begins with a critical stance showing how Christian belief in creation have motivated and justified man's supremacy and power over nature. He argues that this positioning of man, that is often heard in secular discourses in the West, has been linked to environmental destruction and climate change, yet Nasr and many Muslim ecotheologians lack a corresponding critique of Muslim creationism. Fazlhashemi, argues that these thinkers instead develop a critique of western civilization and its economic systems, lifestyles of overconsumption, and also the anthropomorphism and desacralization that are said to characterize Western thinking. Fazlhashemi also raises critical questions concerning how the relationship between science and religion, and between man and nature is perceived in Muslim ecotheological discourses.

In Chap. 4, **Kathrin Winkler** develops some fundamental steps for how postcolonial and indigenous perspectives on sustainability and in religious education can be taken. Winkler argues that it is necessary to create conditions that critically consider how indigenous religions and traditions were delegitimized and consciously suppressed in the course of European colonization and missionization. Winkler also highlights the importance of such steps in cultivating an awareness of the "colonial and missionary societal entanglements of the Christian Western world which played a crucial role in making the current destruction of the ecosystem possible". Referencing the concept of 'solastalgia' which has to do with a sense of lost meaning and belonging, and to Pacific Indigenous and Faith based Re-Storying, the chapter provides an analysis of how religious education rooted in hope can be developed.

The fifth chapter of the Handbook is written by **Ignasi Ribó** who focuses on *jãkatas*, religious parables that play an important role both morally and pedagogically in the Mahayana and Theravada schools of Buddhism. Ribó paints a picture of the complex history of these parables and formulates a critique of attempts to both simplify and overinterpret their meaning. In this chapter, Ribó focuses on two *jãkatas*, in particular *The Tigress* and *Others sow*, which he shows contribute to teaching, and learning about sustainability issues in a higher education setting in Thailand. His research shows how these traditional religious parables address important sustainability perspectives concerning the overcoming of the ideology of humanism, and the cultivation of a posthumanist or post-anthropocentric relationship between humans and nonhumans.

In Chap. 6 **Linda Jonsson** discusses how 'the sustainable classroom' may be conceived as one with a focus on interculturality and diversity. Jonsson describes the possibilities provided by "The Multicultural Almanac", a common resource in Swedish classrooms in the lower grades, to guide both sustainability learning and understanding of different religions. The Almanac is a natural starting point for the school day, and provides an explicit link to the Swedish syllabus for primary school. The Swedish Curriculum states that "Teaching in religion should aim at helping the pupils to develop knowledge of religions and other outlooks on life in their own society and in other parts of the world" (Skolverket, 2022, p. 188). Jonsson shows how using The Multicultural Almanac in Social Studies (which includes Religious Education) classrooms in Sweden can address the Swedish curriculum demands. Taking an intercultural pedagogical and lifeworld-oriented approach, Jonsson shows how The Multicultural Almanac when used as a pedagogical tool can contribute to the pupils seeing themselves as a part of a larger whole.

In Chap. 7, **Hasan Meydan, Recep Kaymakcan** and **Feyza Karaahmetoğlu** develop an analysis of how religiously anchored perspectives on sustainability are integrated in Turkish curricula. The authors present a historical theological background to how present day conceptions and expressions of sustainability are formulated and motivated in this curriculum, and here a framework built around four relevant "broadly Islamic environmental ethics theories" in terms of usefulness, responsibility, virtue and wisdom, is highlighted in order to interpret the curriculum's content and aims. Meydan, Kaymakcan and Karaahmetoğlu finally make a suggestion for how sustainable environmental education could be approached for future curriculum development and preparation of textbooks in schools' religious education.

In Chap. 8, **Harriet Zilliacus**, **Lili-Ann Wolff**, **Pia Mikander** and **Arto Kallioniemi** present the results of an investigation into how the human–nature relationship is introduced into worldview education in the initial comprehensive school years in Finland. These scholars conducted a discourse analysis of Finnish curricula and textbooks in Lutheran and Orthodox religious education; an examination which focused on the position and power of humans in relation to the non-human world. The analysis also looked at how the climate crisis and the need for transformation towards sustainability are taken into consideration in the texts. The authors show that the Finnish curriculum clearly states that humans hold the ethical responsibility for other humans and also non-humans, and aimed to investigate whether there was an absence of human-centredness reflected in the religious education texts. According to the authors, the results of the study raise several critical questions, and the chapter concludes with suggestions on how Christian religious education and the textbooks used need to be changed.

In Chap. 9, **Bodil Liljefors Persson** argues that having knowledge of Indigenous religions and spirituality can deepen understanding of processes of change and permanence, and thus, of sustainability. She takes the case of Maya religion and ritual practices as an example, and discusses how teaching about ecology, sustainability and global ethics and values can benefit from connecting such indigenous religious traditions rituals within Religious Education. The general objective of this discussion is to emphasise the importance of integrating questions of sustainability and ecology in RE classrooms, and to encourage future teachers in RE to develop their knowledge on the relations between religion(s) and sustainability.

Helena Stockinger and **Bernd Ziegler** develop a critical analysis of how "religious Education for Sustainable Development" (rESD) is expressed in German-speaking countries in Chap. 10. Stockinger and Ziegler present some relevant perspectives on what is described as a religious dimension in Education for Sustainable Development, on a normative orientation in rESD and on how references to religious and theological argumentation can play a role in such teaching. Studies showing German-speaking young people's concern about growing climate problems are related to the concept of Design Competence, which aims at having students gain knowledge and act on sustainability issues in an interdisciplinary way. Stockinger and Ziegler, with the help of didactical considerations by thinkers such as Bederna, Gärtner, Birkel and Altmeyer, show a range of possible paths to follow to develop the field of rESD. They also highlight the need for further empirical research that both tests the hypotheses raised in theory on rESD and develop rESD practice.

In Chap. 11, **William Sultmann**, **Janeen Lamb**, **Peter Ivers** and **Mark Craig** report on survey based research where senior secondary school students rated the topics, pedagogies and outcomes of two Religious Education courses offered in Catholic Secondary Schools in Australia. They explain the unique challenge of Catholic Schools in Australia (a sector that educates 18.2% of school age student population), in terms of providing RE that is central to the church's mission and at the same time accommodating a diversity of religious traditions as a secular nation (and education system) requires. The two courses these authors focus on are distinct in that one is considered more traditional (Other RE), the other an innovative 'reconceptualist' approach (RML). They found that the students perceived the topics covered in both courses positively, yet there were significant differences in perceptions of the outcomes and pedagogies. With the RML approach students 'achieved significantly better outcomes… [that] aligned with the Education for Sustainable Development (ESD) goals for learning'.

In Chap. 12, **Bruce Grelle** argues for a rethinking of sustainability education. He makes the case that the concept of sustainability should be better used to critique and ultimately reject that which is '*unsustainable* - the growth-based economic status quo'. He shows us that education which focuses on religions and beliefs – religious literacy education- is critical for the understanding of the current sustainability crises.

In the final chapter (Chap. 13), Olof Franck and Sally Windsor provide concluding remarks about the contributions to the book and suggest paths for further research, starting at the intersection of religion, education and sustainability that will take us into the future.

References

Dasberg, L. (1983). Pedagogy in the shadow of the year 2000. *Phenomenology+Pedagogy*, 117–126.

Franck, O. (2017a). Challenging the teaching of global ethical unity: Religious ethical claims as democratic iterations within sustainability didactics. *Journal of Education for Sustainable Development, 11*(1), 11–18.

Franck, O. (2017b). Highlighting ethics, subjectivity and democratic participation in sustainability education: Challenges and contributions. In O. Franck & C. Osbeck (Eds.), *Ethical literacies and education for sustainable development: Young people, subjectivity and democratic participation* (pp. 1–17). Springer.

Franck, O., & Osbeck, C. (Eds.). (2017). *Ethical literacies and education for sustainable development: Young people, subjectivity and democratic participation*. Springer.

Grimmitt, M., & Read, G. (1975). *Teaching Christianity in RE*. Mayhew.

Hellberg, S., & Knutsson, B. (2018). Sustaining the life-chance divide? Education for sustainable development and the global biopolitical regime. *Critical Studies in Education, 59*(1), 93–107. https://doi.org/10.1080/17508487.2016.1176064

Jickling, B. (1992). Viewpoint: Why I don't want my children to be educated for sustainable development. *The Journal of Environmental Education, 23*(4), 5–8. https://doi.org/10.1080/00958964.1992.9942801

Kaukko, M., Windsor, S., & Reimer, K. (2023). Searching for worlds worth living in. In K. E. Reimer, M. Kaukko, S. Windsor, K. Mahon, & S. Kemmis (Eds.), *Living well in a world worth living in for all: Volume 1: Current practices of social justice, sustainability and wellbeing* (pp. 1–11). Springer Nature. https://doi.org/10.1007/978-981-19-7985-9_1

Kemmis, S. (2023). Education for living well in a world worth living in. In K. E. Reimer, M. Kaukko, S. Windsor, K. Mahon, & S. Kemmis (Eds.), *Living well in a world worth living in for all: Volume 1: Current practices of social justice, sustainability and wellbeing* (pp. 13–26). Springer Nature. https://doi.org/10.1007/978-981-19-7985-9_1

Khaled, L. (2024). New pathway to adolescent wellbeing: The case for online special religious education in public schools. In K. E. Reimer, M. Kaukko, S. Windsor, K. Mahon, & S. Kemmis (Eds.), *Living well in a world worth living in for all. Volume 2: Enacting praxis for a just and sustainable future*. Springer.

Lotz-Sisitka, H. (2004). *Positioning southern African environmental education in a changing context.* Share-Net & Southern African Development Community-Regional Environmental Education Programme.

McDougall, G. (2016). Minority rights and conflict prevention (2010). In *The first United Nations mandate on minority issues* (pp. 334–360). Brill Nijhoff.

Nussbaum, M. C. (2008). Democratic citizenship and the narrative imagination. *Teachers College Record, 110*(13), 143–157.

Osbeck, C. (2017). Discourses of available and sustainable lives: Ethical literacy offered to tweens through fiction Reading. In O. Franck & C. Osbeck (Eds.), *Ethical literacies and education for sustainable development: Young people, subjectivity and democratic participation* (pp. 55–72). Springer.

Pedersen, H., Windsor, S., Knutsson, B., Sanders, D., Wals, A., & Franck, O. (2022). Education for sustainable development in the 'Capitalocene'. *Educational Philosophy and Theory, 54*(3). https://doi.org/10.1080/00131857.2021.1987880

Rothgangel, M., Adam, G., Lachmann, R., Schlag, T., & Schweitzer, F. (2014). *Basics of religious education.* V&R unipress GmbH.

Skolverket. (2022). *Läroplan för grundskolan, förskoleklassen och fritidshemmet, Lgr22* [Curriculum for the compulsory school, preschool class and the leisure-time centre].

UN. (1996). The united nations human rights office of the high commissioner - international covenant on civil and political rights. *General Assembly resolution 2200A (XXI).* Retrieved from https://www.ohchr.org/en/instruments-mechanisms/instruments/international-covenant-civil-and-political-rights

UNESCO. (2022) *What you need to know about education for sustainable development – What is education for sustainable development?* Retrieved from https://www.unesco.org/en/education-sustainable-development/needknow

UNHRC – United Nations Human Rights Commission. (1992). *Declaration on the rights of persons belonging to National or ethnic, religious and linguistic minorities.* https://www.ohchr.org/EN/ProfessionalInterest/Pages/Minorities.aspx

Sally Windsor is an Associate Professor in pedagogical work with a focus on sustainability and international education. Her research and teaching focuses upon teacher training and professional development in terms of interdisciplinary and cross-curricular teaching for social sustainability and global citizenship education.

Olof Franck is Professor in Subject matter education, specializing in the Social Sciences and Associate Professor in Philosophy of Religion at the Department of Pedagogical, Curricular and Professional Studies, University of Gothenburg. His research and teaching interests mainly belong to the fields of religion education, ethics education, ethics, philosophy of religion and religious studies, as well as sustainability education with a focus on social sustainability, ethics and values.

Chapter 2
Truth, Situatedness, and Critique: Toward Religion and Worldview Education in the Anthropocene

Ole Andreas Kvamme

2.1 Introduction

The reflections in this chapter are prompted by the current climate and ecological crises that are having devastating impacts on life on our planet. These crises are, as continuously documented by the UN's climate panel (Intergovernmental Panel on Climate Change, 2018) and nature panel (Intergovernmental Science-Policy Platform on Biodiversity and Ecosystem Services, 2019), the results of the tremendous influence of human activities on the Earth during our age, which is often referred to as the Anthropocene. This term, which was first introduced within Earth system science (Crutzen & Stoermer, 2000), designates a new epoch of geological time – whose current identified starting point is 1945 – dominated by the human impact on our planet. Today, the term Anthropocene is also employed within the social sciences and humanities (Hamilton et al., 2015; Horn & Bergthaller, 2020), moving beyond the question of strictly stratigraphic evidence within Earth science to include the climate system, species extinctions, resource extraction, and waste dumping.

The Anthropocene is also a contentious term (Gough, 2021) that conceals as much as it discloses power asymmetries. The described human impact is expressed in structures and processes involving the unfair distribution of privileges and burdens, reflecting the fact that the responsibility for the current predicament is unequally distributed among the worlds' nations and human populations (Kvamme, 2021). These caveats to an unreflective employment of the concept of the Anthropocene should be acknowledged and included in considerations of ongoing changes.

O. A. Kvamme (✉)
Department of Teacher Education and School Research, The University of Oslo, Oslo, Norway
e-mail: o.a.kvamme@ils.uio.no

S. Windsor, O. Franck (eds.), *Intersections of Religion, Education, and a Sustainable World*, https://doi.org/10.1007/978-3-031-81809-7_2

According to Horn and Bergthaller (2020) the new situation represents a challenge for the humanities, realizing

> that many of the categories used to grasp the relationship between humans and nature have become obsolete.... What is nature when it is fundamentally transformed by human impact? What is culture when it can no longer be understood as a human-made and locally circumscribed environment but has to be seen instead as something that interferes with the forces of nature at a planetary scale? What is humankind, if it is understood as a dominant species whose behavior profoundly affects the Earth system? (p. 5)

Within the context of this paper, we may ask what is the value of school and education, if they have not contributed in the transformation of the conceptions, structures, and practices that have created the conditions for the predicaments the world is currently facing (Orr, 1992).

The following pages explore the potential and resources in religion and worldview education[1] for contributing to such transformations. So far this concern has received moderate interest within religion and worldview education research. An ecological perspective was almost non-existent before the new millennium (Chamberlain, 2000). Martin (2015, p. 34) points out that "a gap remains between our appreciation of the reality of ecological devastation and our work in religious education toward changing the way we think and live." Since Martin's work there has been a growing interest, particularly within catholic religious education, that Altmeyer (2021) suggests may have been encouraged by the encyclical *Laudato Si'*, the call for ecological conversion expressed by Pope Francis (2015). These contributions are embedded in a denominational religious education, framing the research interest within an educational context of education in religious faith. This also includes the substantial contribution of Katrin Bederna (2020), who demonstrates how the challenges involved raise fundamental theological issues and educational concerns alike.

Thus far, there has been a lack of studies that reflect on the possible contributions of integrative religion and worldview education in the Anthropocene. Such an education should neither separate students according to their denominational affiliations nor privilege a particular worldview, be it a Christian or atheistic outlook (Alberts, 2007). Although the issue of privilege is continuously under debate, this is the form of religion and worldview education that is offered as a particular school subject in countries such as Sweden, Norway, England, and Wales, and as one among many alternatives in other countries (e.g., Germany and the Netherlands).

[1] Religious education in various national contexts addresses both religious and non-religious worldviews, also bringing in positions beyond the binary of religious/non-religious (Bråten, 2021). Consequently, scholars have been looking for broader, more inclusive concepts designating the field. I cannot do justice to this discussion here, which is concisely summed up by Valk (2017), who makes a case for "worldview education." I employ "religion and worldview education" – see also Miedema's (2021) "religious and worldview education" and Shaw's conception of "religion and worldview literacy" attentive to Valk's (2017) reminder that religions also have worldviews. "Religious education" is the term used by Alberts, Jackson, and Wright, and is therefore employed when I specifically refer to their accounts.

My ambition is to contribute to reflections on this type of education. Therefore, this paper is theoretical and explorative, revisiting some major and prominent approaches within religion and worldview education during the last 20 years. They are all positioned within an integrative religion and worldview education reflecting increasingly plural European societies. In addition, they all represent comprehensive approaches with a concern for the relationship between school and society, as well as with a conception of education that links educational practices to the formation of the self, in North European didactics expressed in the *Bildung* ideal. First is the critical religious education account developed by Andrew Wright (2007) over a long period of time. This approach is also connected to his concept of religious literacy. Second is the integrative approach coined by Robert Jackson (1997, 2004), which has had a substantial impact on religious education in England and Scandinavia. In the broader European context, it has influenced the work of *Signposts,* the Council of Europe's recommendations for teaching about religions and non-religious worldviews in intercultural education (Jackson, 2014).

Finally, the studies of religion approach to religious education, which was developed by Wanda Alberts (2007, 2008), along with her concept of integrative religious education, is examined. In the various accounts, there is an ongoing conversation between them taking place. Wright and Jackson comment, criticize, and build on each other in their writings, and Alberts brings in both of the other two while establishing her conception of what religious education should be about. While the differences between these approaches should not be minimized, this chapter focuses mainly on the resources they provide for further development, refinement, and critique of religion and worldview education in the Anthropocene.

I identify the vital concerns and formulations of educational purpose that distinguish each of these approaches, mapping what has been prioritized within the research field before I explore the resources in these accounts for religion and worldview education in the Anthropocene. While none of these three accounts addresses ecological and climate challenges as prominent concerns in their conceptions of religion and worldview education, it is possible to identify in their frameworks and approaches entrance points and spaces that pave the way for environmental and sustainability education within religion and worldview education.

2.2 Environmental and Sustainability Education

Thus far, I have situated this paper historically in the Anthropocene, with reference to ecological and climate crises and a concern for increasing global inequality. The UN sustainability agenda (United Nations, 2015) – currently expressed in its 17 sustainable development goals (SDGs) – responds to these predicaments with its vision of sustainable societies, placing education in a key position as SDG 4, and also addresses it in target 4.7 pointing out education as a means to accomplish the other SDGs. Furthermore, the agenda itself addresses the transformation of societal structures and practices, calling for a global learning project involving the various levels of society.

Environmental and sustainability education is distinguished by some common features, which Stevenson, Brody, Dillon, and Wals summarize in five points (Stevenson et al., 2013). First, it is distinctly normative; second, it is interdisciplinary; third, it is participatory and encourages agency; fourth, it transcends the border between formal and informal learning acknowledging students' situatedness in cultural, social, and political contexts outside school; and fifth, it addresses both local and global concerns.

Since the term "sustainable development" was introduced in the context of the UN at the end of the 1980s (World Commission on Environment and Development, 1987), connecting environmental concerns with human development, the sustainability agenda has broadened considerably, addressing the various UN commitments, including the concern for cultural diversity, conflict prevention, and peace. As a consequence, even intercultural education is included in target 4.7 of the UN sustainability agenda, promoting social sustainability. In this paper, with reference to the crises in the Anthropocene, I align with a common approach accentuating the environmental dimension in the notion of sustainability, conceptualized as environmental and sustainability education (see for instance Corcoran et al., 2017). Still, it is also of interest to consider to what extent religion and worldview education in the accounts examined here may be said to address intercultural education, as another aspect of sustainability, intersecting with the other dimensions of sustainability.

2.2.1 The Pursuit of Truth and Cultivation of Truthfulness

A major concern for Andrew Wright (2007) in his account of critical religious education is to reject comprehensive liberalism linked to a worldview wherein capitalism, through its emphasis on the instant gratification of desires, trivializes the human search for meaning. Wright conceptualizes religious education as a search for truth, accentuating an intimate relationship between truth and truthfulness, which here is considered a practice dimension guided by action. With his ideas embedded in critical realism, Wright has a comprehensive notion of truth that includes all reality, emphasizing that the idea of truth is open to a metaphysical dimension. As part of an integrative approach to religious education, Wright operates with an epistemic relativism, stating that a single, total conception of reality is out of reach, thereby producing manifold interpretations. A conspicuous aspect of this kind of religious education is the priority given to traditions (e.g., religious traditions) that maintain and cultivate notions of truth. Truth should not be reduced to singular claims; instead, it should be considered part of larger coherent conceptions that are continuously subject to modification and development.

According to Wright, then, "religious education should enable pupils to wrestle with ultimate truth … given that truth is a deeply contested matter, I am proposing that we should empower pupils to *pursue* truth in an informed and critical manner" (Wright, 2007, p. 7). In this approach, the purpose of religious education is to let students become acquainted with various traditions that offer truth claims, as well as be able to review and select the most convincing tradition, and in this endeavor, also make connections between truth and truthfulness.

> First, religious education must shape and form the whole person, impacting emotionally and intellectually on all they think, say and do. Second, such formation must be guided by, and responsive to, that which is ultimate in the actual order-of-things. Third, pupil should engage actively in their learning, discerning and assimilating truth in a self-consciously responsible manner. (Wright, 2007, p. 78)

By expressing an ambition that goes beyond mere content learning, the practice of religious education might contribute to the development of the personality of the student.

Wright is aware of relationships between religious education and society, and he is concerned with how school might bridge the supposedly broken relationship between truth and the cultivation of truthfulness. This bridging is not solely important for achieving individual development, but it also has bearings on societal well-being. Wright (2007) contends that the pursuit of truth in religious education is key for a society distinguished by diversity:

> A flourishing plural society is one in which different groups can both celebrate their distinctive identities and engage fully in the life of society as a whole. In order to achieve this, a degree of openness and honesty is vital; this requires a willingness and ability to take both difference and commonality seriously. My contention is that the exploration of questions of religious truth, and hence the establishment of appropriate levels of religious literacy, is a basic prerequisite of such a society. (p. 113)

In this way, religious education is woven into one of the great societal challenges of our time: cultivating the well-being of a plural society distinguished by fundamental differences with regard to notions of truth.

Wright's project is explicitly anthropocentric, conceiving truth-seeking as a fundamental aspect of human existence. He is not bothered by concerns for the environment or climate. Still, he is concerned with global injustice, and he opens a space that potentially could also include ecological concerns in his critique of comprehensive liberalism that

> appeals to self-interest as the basic human instinct, reduces human relationships to the level of economic exchange … presides over levels of global inequality that lead directly to the premature and unnecessary deaths of countless men, women and children each day, and perpetuates the self-interest myth that there is no better way of organizing society. (Wright, 2007, p. 29)

A fundamental aspect of this critique is the claim that the ideal of "absolute freedom from constraint is unsustainable, since as relational creatures we are inextricably bound to a complex series of networks and associations" (Wright, 2007, p. 39). This relatedness and connectedness resonate with basic tenets within eco-philosophy (Naess, 1973) and eco-feminism (Merchant, 1996; Warren, 2000), and is central to Wright's conceptions throughout his works and already seen in his *Spiritual Pedagogy*: "Our spiritual lives are determined by the nature of our relationships with ourselves, with other people, with communities and institutions, with the natural world and with the reality – or non-reality – of a transcendent realm" (Wright, 1998, p. 88).

However, in terms of examples and specifications, Wright is primarily engaged with established religious and non-religious worldviews that are to be assessed by students. He introduces five criteria for evaluating worldviews (Wright, 2007, pp. 219–224), emphasizing congruence and coherence, and also bringing in the fertility of the worldview, which refers to its ability to accommodate insights from other knowledge fields. The central example Wright refers to here is contemporary cosmology related to the Christian doctrine of creation. With regard to Christian theology, he could just as well have discussed this point with regard to the fundamental reimagination of basic theological notions within eco-theology (see, for example, McFague, 2000, 2008).

As shown above, Wright (2007) puts truth at the center of his conception of religious education, claiming that "there is an intimate relationship between truth as objective reality and truthfulness as our subjective response to that reality" (p. 14). In its most general form, he states that "a truthful life is one lived in harmony with the way things actually are in the world" (Wright, 2007, p. 15). It is significant that Wright here is in contact with one of the major challenges with regard to the sustainability agenda. For decades, the world has been well aware of the expanding ecological and climate crises, but this insight into *the way things actually are in the world* – to employ Wright's phrase – has not been put into sufficient action. This is why our current sustainability crisis also involves political and pedagogical crises. Following Wright, these crises should be explored in the context of global capitalism, which withdraws the human being from webs of relations and destabilizes the relationship between truth and truthfulness.

2.2.2 Situatedness and Reflexivity

A main concern in Robert Jackson's writings is clarifying and qualifying the contributions of religious education to emerging plural societies in Europe. In his interpretive approach, Jackson (1997, 2004) draws on insights from social anthropology and hermeneutics, accentuating the explorative and interpretive dimensions of religious education. In contrast to Wright's emphasis on congruence and coherence with regard to religious traditions, Jackson points out the dangers of representing religious worldviews as bounded systems of belief, instead emphasizing variety and plurality. Therefore, religions should be studied within specific contexts, and Jackson introduces a three level model for the study of religion: "The interpretation of a religious way of life would thus involve examining the relationship between individuals, relevant groups to which the individual belongs, and the wider religious tradition" (1997, pp. 65–66).

The purpose of religious education, according to Jackson, is "to help children and young people to find their own positions within the key debates about religious plurality" (2004, p. 87). However, equally important as positioning is the interpretive approach as "an approach to understanding the ways of life of others" (Jackson, 2004, p. 87). Here, the contribution from religious education to the well-being of plural societies is distinctly expressed. Religious education is fundamentally to be understood as intercultural education that is crucial to the development of citizenship.

Similar to Wright, Jackson seldom refers to ecological issues or global inequality, and he does not express a concern for climate change. Still, in discussing values education, he holds that

> a vital ingredient of dialogical and interpretive approaches involves pupils in becoming aware of how some of their own prejudices or preconceptions are formed. Both approaches also link the experience of pupils to wider issues, from local to global, opening up issues of human rights and global environmental issues. (Jackson, 2004, p. 140)

Here, Jackson points at how religious education might mediate between the local and the global, a major challenge in environmental and sustainability education, as current crises are global in scope and local in expression (Mannion, 2015).

Central to Jackson's interpretive approach are the interrelated issues of representation, interpretation, and reflexivity. In terms of representation, the concern is who and what are represented, as well as who and what are left out, in the religious education curriculum and classroom practices. With his hermeneutical foothold, Jackson acknowledges that the interpretations of religious ways of life are conditioned and limited by the interpreter's background and preconceptions, which, consequently, must be brought to the fore. Reflexivity follows as the third central issue involving reflections on what bearings our situatedness has on the issues being studied and concerns being raised. This is also a self-critical activity, one in which the teacher and students and, according to Jackson, all who are involved with the representation of religions in society should be encouraged to participate.

Consequently, religious education here emerges as a highly normative and loaded field, one that requires continuous reflexivity in affirming and questioning choices and priorities. This is definitely a concern also relevant to the ability to accommodate and deal with the normative field of environmental and sustainability education. The sustainability agenda, with its various concerns and held together by the notion of justice (Kvamme, 2022a, b), relentlessly foregrounds some concerns and moves other vital concerns to the background.

2.2.3 The Significance of the Academic Field, Impartiality, and Critical-Constructive Didactics

Wanda Alberts is a major proponent of the study-of-religions approach to religious education. In a comprehensive monograph from 2007 (her PhD thesis), Alberts studies integrative religious education in Europe, with a particular focus on England and Sweden. Similar to Wright, Alberts' central reference is the emerging plural societies in Europe, but unlike Wright, who continuously draws on theology, Alberts' foothold is the academic field of studies of religion. Alberts' ambition is to make up for studies of religion's lack of interest in school and religious education. Furthermore, she challenges theology's hegemony and ownership of the field, pointing out how academic conceptions and approaches that are left behind within studies of religion still impact religious education. Key examples are classical phenomenology of religion and the world religions paradigm (Alberts, 2008, pp. 317–318). Moreover, she problematizes how organized religious communities have substantial influence on local curricula in the English context (distinct from the situation in Sweden), challenging institutional impartiality. Alberts is consistently concerned with how the studies-of-religions approach may secure an impartial religious education that does not privilege certain ontological positions, which is important in maintaining liberal societies that live up to human rights standards.

Because the starting point is studies of religion, sometimes it seems that the approach may involve putting context into brackets, developing an ideal position for religious education outside the messiness, disagreements, and compromise of specific contexts:

> Starting from an understanding of the study of religions as an academic discipline, independent from particular institutional contexts and restrictions, we need to develop concepts for education about religion/s independently from external factors: didactics of the study of religions as it should be from an academic point of view, based on research in the study of religions and education. We need to find a position towards didactics within our discipline. (Alberts, 2008, p. 315)

This quote is from Alberts' article calling for *a didactics for the study of religions.* The concept of didactics is taken from the German tradition, with Wolfgang Klafki's (1996) critical and constructive didactics as the central reference. Alberts also formulates the purpose of education within the didactic tradition, and this tradition is definitely sensitive to context. According to Alberts,

> this didactic model aims at promoting (1) *individual self-determination* of life and meaning with respect to e.g. social, vocational and religious matters, (2) *participation* in decision-making about cultural, economic, social and political issues as well as (3) *solidarity* in order to also grant self-determination and participation to others in the face of key challenges that characterize our epoch *(epochaltypische Schlüsselprobleme).* (2007, p. 361)

This is a precise summary of Klafki's account. However, Alberts does not follow up on the central aspect of Klafki's contribution. According to Klafki, the five key problems that he addresses while reconsidering his educational theory in the 1980s "globally interweave our individual and political-societal existence" (1996, p. 154, my translation). These problems are the peace issue, environmental problems, societal inequity within and across nation-states, the hazards and possibilities of information and communication technology, and experiences with love and sexuality. Interestingly, these issues are all addressed in the UN sustainability agenda, making Klafki appear to be an early proponent of sustainability didactics (Kvamme, 2021).

For Klafki, these key issues are attempts to reformulate the main problem in education – what knowledge should be included? – while facing the challenges and conflicts in modern society. Education turns out not to be a question of transmitting certain representations of knowledge but of initiating students into unresolved societal challenges that will determine their future. *Bildung*, the formation of the self, is here tantamount to gaining

> a historically mediated awareness of key problems in current society and – as far as it is predictable – in the future, the insight that everyone is co-responsible for such problems, and a readiness to contribute to the mastery of the problems. (Klafki, 1996, p. 56, my translation)

Alberts does not discuss the challenges that the five key problems constitute for religious education. However, because Klafki's didactic model is decisive for her educational model, this could nevertheless be seen as an invitation to enter into such a discussion.

Furthermore, Alberts acknowledges the critical element in Klafki's account, conceptualizing religious education as critical pedagogy. "In order not to introduce the pupils to unquestioned paradigms, the crucial theoretical and methodological questions need to be addressed in religious education" (Alberts, 2008, p. 319). This is significant for environmental and sustainability education because current imaginaries,[2] structures, and practices have to be questioned, bringing about the transformations necessary for sustainable societies.

As previously mentioned, a major concern for Alberts is considering the implications of updated research in the study of religions for the conceptualization of integrative religious education. When it comes to research contributions, she exemplifies this with postcolonial critique and feminist criticism (Alberts, 2008, p. 317). These fields are in themselves highly relevant to environmental and sustainability education (Gough, 2021; Pashby & Sund, 2020). However, this also demonstrates how priorities within the research agenda impact what is valued in education. Following Alberts' conception, when perspectives from ecocriticism and environmental humanities enter into the study of religion, they also become relevant to the school subject of religious education.

There is a final gateway to sustainability concerns that should be considered with regard to Alberts' account. She emphasizes that plurality of beliefs and values is the starting point for integrative religious education, but she still acknowledges the need to base the subject itself on certain values. These cannot be the values of a particular religion. Instead, she considers the ideological framework of education in general:

> If basic values like human rights, the equality of all human beings or democracy are chosen as a general basis for all school life, what this means for religious education is that all religious traditions will have to be exposed to questions with respect to how their teachings and practice supports or contradicts [*sic*] these values. Furthermore, if criteria and values like democracy or human rights are really taken seriously in integrative RE [religious education], this subject will challenge at a grass-roots level the national and global structures which are often anti-democratic and pay little attention to human rights. (Alberts, 2007, pp. 359–360)

---

[2] The term 'imaginary' makes distinct the social and political aspects of imagination, cf. Castoriadis (1998) and Taylor (2003).

This framework is a part of the model adopted from Klafki, also characterizing Northern European educational systems. In the Norwegian educational context, these values are stated in the objectives clause of primary and secondary school, specified as "respect for human dignity and nature, and on intellectual freedom, compassion, forgiveness, equality and solidarity" (Ministry of Education and Training, 2008). According to Alberts' suggestion, in this national context, one should explore religions and worldviews from the perspective of the values in the objectives clause. Here, these values are "based on fundamental values in Christian and humanist heritage and traditions," (Ministry of Education and Training, 2008) a formulation that scarcely aligns with Alberts' impartial approach. Furthermore, the values "also appear in different religions and beliefs and are rooted in human rights." (Ministry of Education and Training, 2008) What is lacking in this consensus-oriented, harmonious approach is a critical perspective that opens the way for religions and worldviews expressed at various levels in institutional and non-institutional forms to be both possible motivators and hindrances for the promotion of the values in question.

2.3 Discussion: Toward Religion and Worldview Education in the Anthropocene

The approaches to religion and worldview education considered above focus on a plural society made up of manifold religious and non-religious worldviews. The purpose of religion and worldview education is not to educate for a specific worldview but to enable students to navigate within this landscape. Wright presents the approach that most distinctly encourages students to assess and adopt the truth claims that appear to be most convincing, emphasizing the relationship between truth and truthfulness. Jackson connects the purpose of religion and worldview education to learning about the life of the "other" different from oneself. Finally, Alberts puts forward an approach that highlights the importance of impartiality with regard to both content and methods, warranted by a studies-of-religion didactics. Together, these concerns constitute an argument for the pertinence of religion and worldview education with regard to social sustainability, in other words ensuring a society that is able to accommodate diversity and fundamental disagreements built on a foundation of respect across differences.

I have identified concerns for ecological crisis, climate change, and global injustice at the outskirts and fringes of these approaches, all of which were published during the first decade of this century, 20 years after Klafki reimagined his didactic model as a response to new global challenges. The main efforts and achievements within the field during this period have been engaged with reimagining religion and worldview education in a plural society. However, the three approaches together nevertheless present potentials for religion and worldview education in the Anthropocene. Here, I will draw attention to five elements: truth, situatedness, the global dimension, the formative dimension, and the critical and reflexive element.

The first element is the search for truth and truthfulness, which is fundamental to Wright's account. In a broader sense, this educational purpose expresses a conception of humans as beings that search for meaning in their lives. We have seen Wright accentuate the interconnectedness of human life, including fellow human beings, the natural world, and a spiritual dimension. This account builds on a long tradition within religion and worldview education of exploring various conceptions of humankind and worldviews, which are challenged by current crises, calling for critical exploration and reimagination and an openness to existential concerns (Skeie, 2023). It is religion and worldview education as an expression of the humanities that is mobilized here (cf. Horn and Bergthaller's (2020, p. 5) observation quoted previously that the relationship between humans and nature is to be rearticulated in the Anthropocene).

Second is the situatedness of students in specific contexts that become part of the explorations occurring in religion and worldview education practices. This element is most distinctly expressed in Jackson's interpretive approach, which encourages education that transcends the classroom and positions religious life in the context of the local community. In religion and worldview education, excursions to various religion and worldview communities exemplify this aspect, which is also encouraged in environmental and sustainability education. It is significant that these activities establish learning content that is not detached from the lives of students; instead, it demonstrates the plurality of the communities to which these students already belong. This conceptualization of situatedness is key in our age of climate and ecological crises.

The global aspect of religion and worldview education is touched upon by Jackson (2004). Educational practices often explores phenomena that are global in terms of their expression, representing great diversity within the various traditions. Religion and worldview education is cosmopolitical in the sense that students become familiar with narratives, symbols, and imaginaries reflecting experiences from a variety of contexts worldwide. This global outlook may also be an invitation to imagine the situatedness of the distant other and, at the same time, defamiliarize the structures and practices in which students are embedded. While the global dimension in environmental and sustainability education is often difficult to express (Mannion, 2015), religion and worldview education provides resources that can be employed for this purpose.

The fourth element is the formative dimension in religion and worldview education that involves and transcends the development of knowledge, skills, and attitudes. Wright hopes students will be able to live a truthful life in harmony with the reality of the world. Jackson accentuates how an understanding of the other may also lead to a new understanding of oneself. Alberts introduces the didactic model of Klafki, calling for an emancipatory religious education enabling self-determination, co-determination, and solidarity. At the same time, all three authors are aware of the pedagogical paradox involved. Emancipatory and impartial religion and worldview education should not force students into any position, be it religious or non-religious, but it should facilitate an openness that promotes the search for meaning. Similarly, environmental and sustainability education should not force students into environmental awareness and solidarity; instead, it should facilitate educational contexts wherein such formation is promoted. Alberts' suggestion to let educational and societal values guide the study of religion is intriguing. The dynamics of the *Bildung* ideal are brought into play, involving society, students, and educational practices. In a time when respect for all (not just human) life, global justice, future generations, and cultural diversity are increasingly evoked in educational documents (e.g., Ministry of Education, 2017; United Nations Educational, Scientific and Cultural Organization (UNESCO), 2015), this proposal is particularly pertinent.

Finally, when Alberts adopts Klafki's didactic model and combines it with her studies-of-religion approach, she introduces a critical perspective into religion and worldview education, continuously questioning power relations and the preconditions of conceptions and structures maintaining religion and worldview education itself and being subject to exploration by students. We have seen similar critical perspectives expressed in the other two accounts, by Jackson particularly linked to practicing representation and reflexivity. The critical perspective is also present within environmental and sustainability education, something that is decisive for enabling necessary transformations into sustainable societies. Given this reflexive nature, it even calls for reflections on how the ecological crisis, climate crisis, and increasing global inequity so far have been largely invisible within the field of integrative religion and worldview education.

2.4 Concluding Remarks

The aim of this chapter has been to explore how some major approaches to religion and worldview education offer directions for contributing to the interdisciplinary field of environmental and sustainability education, sketching out a contextually situated, existential, global-oriented, formative, and reflexive religion and worldview education.

Here, at the end, I would like to point out a few aspects that call for particular attention. One is the position of the *Bildung* ideal as an expression of the formative element, which is central to Alberts in the educational grounding of her approach, with reference to Klafki (1996). In this chapter, I have acknowledged Klafki's renewal of educational theory as significant in the Anthropocene. However, it should be mentioned that the imaginary of the human that emerges in Klafki's writings is predominantly an expression of modernity focusing on autonomy, agency, and human control over nature (Kvamme, 2021). Action is needed, but a sole emphasis on agency and control may also be conceived of as part of the problem, bringing ambiguity into this imaginary (Lysgaard & Bengtsson, 2022). Other imaginaries, such as cautiousness (Paulsen, 2021) and vulnerability (Kvamme, 2021) should be searched for, oriented towards human situatedness in a web of relationships (Benhabib, 1992; cf. Wright, 2007). In religion and worldview education, such a reimagining can be carried out in the explorations of the myriad of narratives, symbols, and imaginaries that constitute this school subject.

Another conspicuous aspect that should be mentioned is the lack of attention to ethics and ethics education (also called moral education) within the approaches of Jackson, Alberts, and Wright (although Wright's concept of truthfulness includes ethical considerations). This is not surprising as it confirms the impression that ethics education and religion and worldview education as academic disciplines to a large extent have developed as distinct fields with limited contact (Kvamme, 2017). Problems of this divide become specific when ethics is part of school subjects (e.g., in Norway (Kvamme, 2017), Sweden (Osbeck & Skeie, 2014), Iceland (Gunnarsson, 2014), and Scotland (Conroy, 2014)). With regard to the focus of this chapter, the challenges faced in the Anthropocene are certainly both ethical and political (Kvamme, 2022a, b; Sporre et al., 2022), calling for mediations between ethics education and religion and worldview education.

Decisive for religion and worldview education is the focus on meaning-making, letting students' experiences and existential concerns emerge and be discussed in the educational practices of the school subject. In our age, there is a need for educational spaces that accommodate students' anxieties, worries, and hopes with regard to the global crises caused by human activities. Religion and worldview education can offer such spaces.

Acknowledgments The author acknowledges helpful comments from professor Oddrun Marie Hovde Bråten, The Norwegian University of Science and Technology to a draft of this chapter presented 20 April 2023 at a research seminar facilitated by the Swedish and Norwegian religious education networks and hosted by University of Gothenburg.

References

Alberts, W. (2007). *Integrative religious education in Europe. A study-of-religions approach.* Walter de Gruyter.

Alberts, W. (2008). Didactics of the study of religions. *Numen, 55*(2–3), 300–334.

Altmeyer, S. (2021). Religious education for ecological sustainability: An initial reality check using the example of everyday decision-making. *Journal of Religious Education, 69*(1), 57–74. https://doi.org/10.1007/s40839-020-00131-5

Bederna, K. (2020). *Every day for future: Theologie und religiöse Bildung für nachhaltige Entwicklung* (2nd ed.). Grünewald.

Benhabib, S. (1992). *Situating the self: Gender, community, and postmodernism in contemporary ethics*. Routledge.

Bråten, O. (2021). Non-binary worldviews in education. *British Journal of Religious Education, 44*(3), 325–335. https://doi.org/10.1080/01416200.2021.1901653

Castoriadis, C. (1998). *The imaginary institution of society*. MIT Press.

Chamberlain, G. L. (2000). Ecology and religious education. *Religious Education, 95*(2), 134–150.

Conroy, J. C. (2014). Religious education at schools in Scotland. In M. Rothgangel, R. Jackson, & M. Jäggle (Eds.), *Religious education at schools in Europe. Part 2: Western Europe* (pp. 233–260). Vienna University Press.

Corcoran, P. B., Weakland, J. P., & Wals, A. E. J. (Eds.). (2017). *Envisioning futures for environmental and sustainability education*. Wageningen Academic Publishers.

Crutzen, P. J., & Stoermer, E. F. (2000). The Anthropocene: An epoch of our making. *Global Change Newsletter, 41*(May), 17–18.

Francis. (2015). *Laudato Si': On care for our common home*. Encyclical.
Gough, A. (2021). Education in the Anthropocene. In C. Mayo (Ed.), *Oxford encyclopedia of gender and sexuality in education*. Oxford University Press. https://doi.org/10.1093/acrefore/9780190264093.013.1391
Gunnarsson, G. J. (2014). Religious education at schools in Iceland. In M. Rothgangel, G. Skeie, & M. Jäggle (Eds.), *Religious education at schools in Europe. Part 3: Northern Europe* (pp. 127–150). Vienna University Press.
Hamilton, C., Bonneuil, C., & Gemenne, F. (Eds.). (2015). *The Anthropocene and the global environmental crisis. Rethinking modernity in a new epoch*. Routledge.
Horn, E., & Bergthaller, H. (2020). The Anthropocene. Key issues for the humanities. .
Intergovernmental Panel on Climate Change. (2018). *Global warming of 1.5°C. Special report*. https://www.ipcc.ch/sr15/
Intergovernmental Science-Policy Platform on Biodiversity and Ecosystem Services. (2019). *Summary for policymakers of the global assessment report on biodiversity and ecosystem services*. https://www.ipbes.net/document-library-catalogue/summary-policymakers-global-assessment-laid-out
Jackson, R. (1997). *Religious education. An interpretive approach*. Hodder & Stoughton.
Jackson, R. (2004). *Rethinking religious education and plurality: Issues in diversity a pedagogy*. Routledge.
Jackson, R. (2014). *Signposts: Policy and practice for teaching about religions and non-religious world views in intercultural education*. Council of Europe.
Klafki, W. (1996). *Neue Studien zur Bildungstheorie und Didaktik. Zeitgemäβe AllgemeinBildung und kritisch-konstruktive Didaktik*. Beltz Verlag.
Kvamme, O. (2017). The significance of context: Moral education and religious education facing the challenge of sustainability. *Discourse and Communication for Sustainable Education, 8*(2), 24–37. https://doi.org/10.1515/dcse-2017-0013
Kvamme, O. (2021). Rethinking Bildung in the Anthropocene: The case of Wolfgang Klafki. *HTS Teologiese Studies/Theological Studies, 77*(3), a6807. https://doi.org/10.4102/HTS.V77I3.6807
Kvamme, O. (2022a). Ethical grounding of critical place-based education in the Anthropocene. In M. Paulsen, J. Jagodzinski, & S. M. Hawke (Eds.), *Pedagogy in the Anthropocene: Re-wilding education for a new earth* (pp. 277–294). Palgrave Macmillan. https://doi.org/10.1007/978-3-030-90980-2_14
Kvamme, O. (2022b). Facets of justice in education: A petroleum nation addressing United Nations sustainable development agenda. *Ethics and Education, 17*(2), 163–182.
Lysgaard, J. A., & Bengtsson, S. L. (2022). Action incontinence: Action and competence in dark pedagogy. In M. Paulsen, J. Jagodzinski, & S. M. Hawke (Eds.), *Pedagogy in the Anthropocene: Re-wilding education for a new earth* (pp. 107–128). Palgrave Macmillan. https://doi.org/10.1007/978-3-030-90980-2_6
Mannion, G. (2015). Towards glocal pedagogies: Some risks associated with education for global citizenship and how glocal pedagogies might avoid them. In J. Friedman, V. Haverkate, B. Oomen, E. Park, & M. Sklad (Eds.), *Going glocal in higher education: The theory, teaching and measurement of global citizenship* (pp. 19–34). University College Roosevelt.
Martin, M. K. (2015). The cry of earth and conflict with human cultures: A reflection for Christian religious educators. *Journal of Religious Education, 63*(1), 25–36.
McFague, S. (2000). *Life abundant: Rethinking theology and economy for a planet in peril*. Fortress Press.
McFague, S. (2008). *A new climate for theology: God, the world, and global warming*. Fortress Press.
Merchant, C. (1996). *Earthcare: Women and the environment*. Routledge.
Miedema, S. (2021). On relevant knowledge and valorization in religious and worldview education. *Zeitschrift für Pädagogik und Theologie, 73*(1), 5–15. https://doi.org/10.1515/zpt-2021-0002
Ministry of Education and Research. (2008). *Education act*. https://www.regjeringen.no/en/dokumenter/education-act/id213315/?regj_oss=10

Ministry of Education and Research. (2017). *The core curriculum. Values and principles for the primary and secondary school.* https://www.regjeringen.no/en/dokumenter/verdier-og-prinsipper-for-grunnopplaringen%2D%2D-overordnet-del-av-lareplanverket/id2570003/
Naess, A. (1973). The shallow and the deep, long-range ecology movement: A summary. *Inquiry, 16*(1), 95–100.
Orr, D. (1992). *Ecological literacy: Education and the transition to a postmodern world.* State University of New York Press.
Osbeck, C., & Skeie, G. (2014). Religious education at schools in Sweden. In M. Rothgangel, G. Skeie, & M. Jäggle (Eds.), *Religious education at schools in Europe. Part 3: Northern Europe* (pp. 237–266). Vienna University Press.
Pashby, K., & Sund, L. (2020). Decolonial options and foreclosures for global citizenship education and education for sustainable development. *Nordic Journal of Comparative and International Education (NJCIE), 4*(1), 66–83. https://doi.org/10.7577/njcie.3554
Paulsen, M. (2021). Cautiousness as a new pedagogical ideal in the Anthropocene. In K. B. Petersen, K. von Brömssen, G. H. Jacobsen, J. Garsdal, M. Paulsen, & O. Koefoed (Eds.), *Rethinking education in light of global challenges* (pp. 220–232). Routledge. https://doi.org/10.4324/9781003217213
Shaw, M. (2020). Towards a religiously literate curriculum – Religion and worldview literacy as an educational model. *Journal of Beliefs & Values, 41*(2), 150–161. https://doi.org/10.1080/13617672.2019.1664876
Skeie, G. (2023). Etiske verdier og eksistensiell utforskning: Hvordan kan skolens religions- og livssynsundervisning arbeide med bærekraftig utvikling som tverrfaglig tema? [Ethical values and existential exploration: How can religious and worldview education in school address sustainable development as an interdisciplinary topic?] *Nordisk tidsskrift for pedagogikk og kritikk, 9.* https://doi.org/10.23865/ntpk.v9.4039
Sporre, K., Lotz-Sisitka, H., & Osbeck, C. (2022). Taking the moral authorship of children and youth seriously in times of the Anthropocene. *Ethics and education, 17*(1), 101–116. https://doi.org/10.1080/17449642.2021.2024991
Stevenson, R. B., Wals, A. E. J., Dillon, J., & Brody, M. (2013). Introduction. In R. B. Stevenson, M. Brody, J. Dillon, & A. E. J. Wals (Eds.), *International handbook of research on environmental education* (pp. 1–7). Routledge.
Taylor, C. (2003). *Modern social imaginaries.* Duke University Press.
United Nations Educational, Scientific and Cultural Organization (UNESCO). (2015). *Rethinking education: Towards a global common good?* https://unevoc.unesco.org/e-forum/RethinkingEducation.pdf
United Nations General Assembly. (2015). *Resolution 70/1. Transforming our world: The 2030 agenda for sustainable development.* https://www.un.org/en/development/desa/population/migration/generalassembly/docs/globalcompact/A_RES_70_1_E.pdf
Valk, J. (2017). Worldview education. In D. R. Wielzen & I. Ter Avest (Eds.), *Interfaith education for all* (pp. 227–242). Sense Publishers.
Warren, K. J. (2000). *Ecofeminist philosophy: A Western perspective on what it is and why it matters.* Rowman & Littlefield, Publishers, Inc.
World Commission on Environment and Development. (1987). *Our common future.* https://sustainabledevelopment.un.org/content/documents/5987our-commonfuture.pdf
Wright, A. (1998). *Spiritual pedagogy: A survey, critique and reconstruction of contemporary spiritual education in England and Wales.* Culham College Institute.
Wright, A. (2007). *Critical religious education, multiculturalism and the pursuit of truth.* University of Wales Press.

Ole Andreas Kvamme is Associate Professor at Department of Teacher Education and School Research, University of Oslo, Norway. In his research he explores the ethical-political dimension of environmental and sustainability education, including perspectives from critical cosmopolitanism and environmental ethics, also exploring contributions from religion and worldview education.

Chapter 3
Islamic Ecotheology

Mohammad Fazlhashemi

3.1 Introduction

The intensifying debate over environmental destruction, global heating, and humanity's roll in the emission of that greenhouse gases threating the Earth, humanity, and all animal and plant life, has been accompanied by a parallel debate over these same issues by Muslim thinkers. The reason is that these threats are not limited to countries in the West but undermine the conditions for life on the entire planet.

One element of the debate in the West has been the role played by religion, or, more precisely, Christianity in the environmental and climate crisis. The Christian creation story, which elevates human beings to lords of creation with sovereignty over the Earth, has been a prominent target of criticism. Its perspective has been accused of conferring legitimacy upon humanity's plundering of nature. A similar viewpoint can be found in the Muslim creation narrative, which presents human beings as the supreme creation in whose service God has created everything under the heavens. For Muslims, the question is how to engage with this criticism. Do Muslim theologians agree that Islam's creation narrative and view of humanity have contributed to the overexploitation of the resources of nature? How do Muslim theologians regard Islamic creation theology's view of the relation between humanity and God's non-human creations?

This debate is connected to a larger question that relates to the role of religion in climate and environmental issues. Do Muslim theologians agree that Islam is a part of the problem, or do they present Islamic creation theology and its theological ethics as the solution to the problem? Does Islamic ecotheology see the solution as lying in religion/Islam or somewhere else? Do they think that the solution lies in the individual or should the issue be dealt with at the collective level as politics?

M. Fazlhashemi (✉)
Department of Theology, Uppsala University, Uppsala, Sweden
e-mail: mohammad.fazlhashemi@teol.uu.se

S. Windsor, O. Franck (eds.), *Intersections of Religion, Education, and a Sustainable World*, https://doi.org/10.1007/978-3-031-81809-7_3

It is difficult to find support for the view that the Islamic creation theology's view of humanity and its conviction that God created the natural world to serve humanity as a contributory cause of the environmental and climate crises. Instead, some Muslim theologians have developed a civilizational critique of the West in which the blame is put on European/Western anthropocentrism, individualism, Cartesian deism, industrialism, colonialism, the capitalist economic system, and, last but not least, the European model's claim to universalism. These are identified as the most important factors, which have together laid the foundations for humanity's destructive impact on the environment and climate. It is important to emphasize that when Muslim theologians talk about of humanity's guilt and responsibility for the environmental crisis, they are referring specifically to those in the West. Along with the Western worldview and its economic system, that is to say, capitalism and its inexhaustible pursuit of profit, they regard the Western lifestyle and overconsumption as the fundamental causal factors behind the release of greenhouse gases, global heating, environmental destruction, and the overexploitation of natural resources.

This critique of civilization and the identification of European/Western actors as responsible for the negative developments mean that the responsibility for taking steps to address the environmental issue is primarily the responsibility of states in the West. At the same time, Muslim theologians exonerate Islamic teachings from any responsibility for the aforementioned problems. A question arises in this context: what theological arguments do Muslim theologians advance when rejecting criticisms of Islam's creation theology? A second question is: what line of argument do they adopt with regard to Muslims' embracing of the Western lifestyle and its culture of overconsumption as well as industrialization and modernization more broadly?

3.2 The Anthropocentric Perspectives and Desacralization

In this article I focus on the ideas of the Iranian philosopher Seyyed Hossein Nasr (b. 1933), who has been an influential figure in this debate. He identifies the background to the problems of global heating and environmental destruction in the paradigm shift that took place in Europe as a result of the Renaissance and subsequent developments. As he sees it, a paradigm shift took place in Europe during the Renaissance that paved the way for the environmental and climate crises of the present day. It was at that time that the theocentric worldview was replaced by an anthropocentric viewpoint. As a result of this paradigm shift, humanity was given a central position and its status elevated to that of God. Since the fifteenth century, this new worldview has displaced God from his supreme position in creation. In practice, humanity was elevated to the status of God. According to Nasr, this worldview paved the way for humanity's mastering, taming, pillaging, and exploitation of the resources of the natural world (Nasr, 2007).

With the scientific revolution, this anthropocentric worldview attained a new primacy, and God, revelation, and the religious worldview were made subordinate to natural science. The next step, which completed this process, was taken by the French philosopher René Descartes (1596–1650), who divided reality into thought/spirit and material. The Cartesian worldview no longer treated the world as an organism but rather as a machine, a gigantic clockwork apparatus that God, the watchmaker, had set in motion at the beginning of time. The task of disciplinary knowledge, or, more precisely, the natural sciences, was to discover the rules by which this colossal machine worked. Descartes's division of reality into two fundamentally separate domains paved the way for the plundering of nature in the modern era, argue Nasr and many other Muslim thinkers. Nature was regarded as something for humanity to dominate, defeat, and exploit as it pleased. Another task for the natural sciences was to map out the resources of nature, which was treated as an inexhaustible storehouse. Humanity viewed itself as entitled to exploit, or, more precisely, use for its own benefit everything that this endless store had to offer, including: the felling of forests; the rampant mining of the earth, which left irreversible scars upon nature and caused profound damage to the ecosystem; and the plundering of the resources of earth and sea in such a way that entire species were made extinct and ecosystems collapsed. This exploitation had a highly negative impact on the environment and climate, which, though presumably not intentional, was a consequence of prioritizing humanity's rights and interests above all else. Accentuating this development was the fact that the mechanistic world view did not regard itself as bound by any ethical limits.

One consequence of the Cartesian worldview was that faith and God became superfluous (Nasr, 1993). Its worldview regarded humanity as a worldly creation that needs only to follow a soulless rationalism. Faith in spiritually impoverished reason meant that humanity could doubt and call into question given metaphysical truths. This meant that humanity did not need to feel constrained by truths of an ethical or metaphysical character. In a world where such considerations no longer troubled the human conscience, nothing could prevent humanity from dominating, taming, and exploiting the world and nature as it wished (Nasr, 2007).

Nasr concludes that today's environmental and climate crises have their basis in a philosophical, theological, and ethical crisis. It is a spiritual crisis that rests upon the rejection of religion as something that has nothing to contribute to the solution of the current crisis. As Nasr sees it, humanity has failed to understand God's purpose with creation and its own responsibility to God. He rejects the notion that humanity can deal with the crisis solely by means of science. His main point is that religion can equip humanity with an ethical basis for its relationship to nature. Religion's most important contribution is to resacralize nature and provide humanity with theological ethics, according to Nasr. The latter equips humanity with an ethical foundation and a moral compass that will help it to curb the plunderer's mentality with regard to nature (Nasr, 2007).

Theological ethics and the sacralization of nature stand in opposition to the modern scientific tradition that is dominated by logical positivism and naturalism. This scientific perspective has led to a secularization of the modern worldview that bears much of the blame for today's environmental and climate crises. Contradicting this view, the Muslim perspective on existence is taken to proceed from a transcendent reality and the existence of an intellect behind human reason. Nasr refers to unchangeable ahistorical principles and a metaphysical truth that is not historically, socially, or culturally conditioned. He underscores that modern Western science has capitulated to experimental and analytical methods and for this reason is unable to learn from such principles.

The desacralization of nature is only partly grounded on the Cartesian worldview. Criticism is also directed at Christian theology's relationship to modern science. Christian theologians, Nasr argues, have felt threated by the burgeoning scientific worldview and the progress of science. Their way of countering this threat was to use science in the service of faith. Instead of disarming the threat, they found themselves in a complicated relationship to science and became dependent on it. The Christian worldview that was based on natural theology was shaken to its core by the destruction of earlier scientific ideas that followed upon new scientific discoveries. According to Nasr, this development has its origins in the Averroestic philosophical school, which laid the foundations of a Christian natural theology that paved the way for a desacralization of the world and nature.

The desacralization of nature reached its zenith with the scientific revolution. From this point on, nature would be treated from a secular perspective, which further weakened the standing of faith. The new scientific worldview, which was secular and material in character, rejected metaphysical truths. This led instead to a kind of alienation from nature (Nasr, 1993).

The scientific revolution led to enormous advances in the realm of the natural sciences. Trust in science, reason, and, above all, Newtonian physics knew no limits. Science was used as a screen through which nature was viewed. Everything associated with metaphysics was jettisoned. Empirically based science was regarded as having the answer to everything. The task of science was to discover the laws and rules that governed nature in order to reveal its secrets. This would, in the words of Francis Bacon (1561–1626), help humanity to master nature and thereby acquire power and create prosperity (Nasr, 2001).

This view of science, in combination with the division of reality into two separate worlds—theology/philosophy on the one hand, and science on the other—resulted in the loss of an integrated picture of reality and an organic understanding of nature. This division led in turn to the belief that God, who had originally set the machinery of nature in motion, no longer had anything to do with it. Nature continued to operate its almost unvarying mechanism. The mechanistic worldview ultimately resulted in a perception of nature as a masterless machine that needed to be tamed, defeated, subordinated to humanity's will, and placed at the service of humanity.

This division drove revelation, its norms, and its ethical compass to the periphery. Arguments and explanations grounded on religion were marginalized, religious norms and values were viewed as irrelevant to humanity's relationship to nature and its handling of and attitudes towards nature's resources. Its relationship to nature was instead defined by the profit motive. Nature was regarded as an unlimited body of resources that humanity needed only to find new technological ways of using and exploiting, argues Nasr.

3.3 Homo Economicus

Another components in the Muslim ecotheological debate over the causes of the current environmental and climate crisis is the libertarian perspective. Its problems arise from humanity's behaviour and view of itself as *Homo Economicus*. It is a viewpoint that treats human beings as amoral entities whose primary motivation or driving force is self-interest, self-realization, and the pursuit of economic profit. Nasr sees its pursuit of profit as being of a piece with naiveté towards the Earth's resources as inexhaustible and eternal and the belief that these resources can be exploited relentlessly in order to achieve maximum profits. For him, *Homo Economicus* is seen as operative within the framework of the modern economy, which is obsessed with accumulating profit. This is a viewpoint that provides the basis for ruthless exploitation of the world's resources, something that in turn has led to the environmental destruction and emission of greenhouse gases that have had an extremely negative impact on the climate. This critique has its point of departure in a theological perspective that opposes the libertarian worldview because of the instrumentality of its core principles. This critique is developed into a conflict of paradigms that counterposes the spiritual perspective to the instrumental. It is a conflict that has far-reaching consequences for the environment and the climate.

Nasr identifies thinkers such as David Hume (1711–1776) and Jeremy Bentham (1748–1832) as the central figures for profit-based thinking and the emergence of the view of human beings as *Homo Economicus*. On this view, the core of the problem is that it accords a central place to the value that everything has for humanity. Nasr accuses this viewpoint of not paying attention to nature, the climate, and the world's resources and their state (Farahanifar & Farahanifar, 2013).

Another problem is the libertarian perspective that puts human beings' individual interests ahead of everything else. The ethical system that drew on modern European perspectives suppressed religious and metaphysical conceptual models and elevated human beings to the status of absolute and supreme lord of nature and the environment. This intellectual lineage is primarily characterized by how it allowed itself to be blinded by the giant steps made by science even as it left no place for God and subordinated all aspects of existence to humanity's own interests and freedom of action (Farahanifar & Farahanifar, 2013).

3.4 Nature as God's Sign

Islamic creation theology has been held up as a counterweight to the Cartesian worldview, because it does not differentiate between the material and the immaterial. Nor does it regard nature as a gigantic machine to be dominated and exploited. Rather, nature is treated as an arena for God's sign, *Āyât*. Studying God's sign in nature provides insights into nature's complex rules and, with it, the omnipotence of God/the Creator. Science is seen as being not in competition with revelation but rather as a tool that helps to lay bare the complexity of creation. It discovers and documents the laws governing nature. It is a mode of knowledge that helps human beings to achieve insight into the need for an all-powerful designer/creator.

Like creation theology in the Abrahamic tradition, God is seen as the Almighty Creator who has created everything and put everything on earth at the service of humanity (Qur'an, 67:15 & 22:65). The question that arises is whether the Qur'an's assertion that God has created everything to serve humanity could be interpreted as meaning that God is giving humanity carte blanche to rule over the earth and all non-human creation and that humanity has the right to engage in plundering nature. Muslim theologians including Nasr reject this charge by arguing that humanity, despite its position as the apex of creation, should not be seen as different in kind from God's non-human creation. They underscore that God created humanity from the same materials as other, non-human creation. All are made from the same building blocks, the argument goes. This common denominator is seen as proof of humanity's affinity with the natural world. This conclusion is based on statements in the Qur'an that everything owes its existence to God, whose creative power has created and continues to sustain all of existence. This and similar statements in the Qur'an are taken as arguments for why humanity should not be accorded a privileged status above nature.

In order to underscore this point, it is argued that Islamic creation theology treats God as the only rightful owner of the world since God is its sole creator (Qur'an, 2:115 & 3:189).[1]

This is understood to mean that humanity cannot lay claim to own God's creation, which in turn means that humanity cannot make use of something that does not belong to it. God's status as the only rightful owner means that humanity cannot treat nature as it pleases or for its own interests, regardless of whether those interests are profit or something else.

Islamic creation theology serves as a foundation for Muslim ecotheologians in their arguments against humanity's supremacy over God's non-human creation. Its most important constitutive elements include the following:

- The testimony of all creation, *Shahada*, to God's greatness and omnipotence.
- Humanity's role as God's sign, *Āya, Vestigia Dei*.
- Humanity's role as God's servant, *ʿAbd*.
- Custodial responsibility, *Amana*.
- Humanity's role as God's representative, *Khalifa*, on earth.
- Islamic ethic's emphasis upon moderation in consumption.

Testimony, *Shahada*, is presented as proof that humanity is not the only witness to God's greatness and omnipotence. It is emphasized that God's non-human creations, such as the sun, moon, stars, the wind, animals, and plants, bear witness to

---

[1] These verses are highlighted as evidence that it is God who is the sole owner of creation:

> And God's is the east and the west; therefore wherever you turn, there is the face of God. Qur'an, 2:115 and And to God belongs the kingdom of the heavens and the earth; and God has power over all things. Qur'an, 3:189.

the same thing. This is interpreted as saying that Islam's creation theology does not make a sharp distinction between human and non-human creation. To this can be added the fact that non-human creation is as important as humanity insofar as both are regarded as God's signs (Nasr, 1993).

> Muslim sages who referred to the cosmic or ontological Qur'an ... they saw upon the face of every creature letters and words from the cosmic Qur'an ... they remained fully aware of the fact that the Qur'an refers to phenomena of nature and events within the soul of man as āyāt ... for them forms a nature were literally Āyāt Allah (Nasr, 1993, pp. 130–131).

One consequence of this conception of non-human creation/nature as something that is not completely different in kind from humanity, but rather a part of the world alongside human beings, is that it is not regarded as an enemy to be defeated or something for humanity to rule over. Like human beings, non-human creations are regarded as God's signs, *Āyāt, Vestigia Dei*. By presenting humanity as God's sign, *Āya*, humanity is integrated into all God's creation. In this way, it becomes merely a part of creation, a part of nature—not its master.

Another reason why Islamic creation theology does not give humanity free rein to plunder nature is its view of humanity as God's servant, *ʿAbd*. From this perspective, humanity, as a good servant to God, is expected to take hand of that which God has created "in the best of forms" and that humanity should take care of this creation to the best of its abilities. This is also something that humanity has born witness to in connection with creation, Nasr writes. Seen from this perspective, testimony, *Shahada*, means that humanity is fully aware of its responsibility (Qur'an 7:172).

At the same time, Nasr presents humanity as having been given a particular responsibility among all of God's creations to safeguard its continued existence. The Qur'an here uses the concept of *Amana*, which has been interpreted as meaning that God temporarily lends his creation to humanity and that humanity accepts custodial responsibility.[2] Humanity is also described as the part of God's creation that is best able to achieve the standing of God's representative, *Khalifa*, on earth.It is this verse from the Qur'an that is considered to strengthen this view:

> (O Muhammad) when your Lord said to the angels: "Verily, I am going to appoint a vicegerent, *Khalifa*, in the earth." They said: "Will You (O our Lord) place one in it who will make mischief and shed blood, while we celebrate Your praise and extol Your holiness? He said: "Verily, I know what you do not know." (Qur'an, 2:30).

Custodial responsibility, *Amana*, is presented as an important foundation of this theological conception. It is seen as counteracting the belief that humanity has been given the opportunity or permission to rule over the Earth and nature without any limits whatsoever. Humanity's awareness of this responsibility works proactively in that it is regarded as providing the insight that God is the creator of the world and

[2] The idea of custodial responsibility in Islamic creation theology is based on this verse.

> Verily We offered the trust, *Amana*, to the heavens and the earth and the mountains, but they refused to bear it and were afraid of it; but man undertook it; verily he was unjust, ignorant. Qur'an, 33:72.

that humanity is only one of God's creations, albeit one that has been given the responsibility of administering and taking care of the Earth/nature and its resources during its earthly existence. Recurrent references to verses in the Qur'an follow a recurrent tradition in Muslim theological debate. References to Qur'anic verses are the ultimate proof that the interpretation being advocated is solidly grounded in revelation.

Humanity's role as God's representative on earth, *Khalifa*, is closely bound up with the notion of custodianship and means that the administration of God's creation is not humanity's only task. Another idea that is connected with this notion is that humanity is in a relationship of dependency to creation. The idea of custodianship takes its point of departure in the fact that humanity's relation to God's creation is a mutual relationship between creation and humanity. This idea is regarded as countering the view of nature as an inexhaustible storehouse that humanity can use up as it likes. Custodianship means that human beings do not own each other, the Earth, animals, trees, and everything else that forms part of creation. It merely confirms that humanity is a part of creation. The only thing that differentiates it is that upon its creation it was given the task/responsibility of administering God's creation in the best way possible. One consequence of the mission of custodianship is that humanity is responsible to God, who entrusted it with this responsibility. This is seen, in turn, as meaning that because humanity is responsible to God, it cannot treat nature with abandon.

Humanity's custodial responsibility and its dependence on nature are seen as a precondition for creating balance in the world. This goes back to the idea that it is not merely the one who gave humanity its mission who will hold it accountable for its actions towards nature and the environment. Invoking the Qur'an, Islamic eco-theologians argue that if humanity fails in its responsibility and mismanages nature/the environment, the latter will strike back with full force. Climate change and global heating offer examples of how nature will strike back with devastating effects when humanity mismanages its mission. On the other side, if humanity choose to live in harmony with nature and treat it well, nature will provide generously and in a way that enriches human life.

Nasr's views on this issue are of course not shared by all Muslim theologians. As I understand it, Islamic theology regards good actions in such matters as having intrinsic value. This does not necessarily contradict Nasr's notion of good actions as having good consequences, but it may also be the case that the consequences of good actions extend beyond what it immediately visible.

The notion of custodianship can also be seen as a counterweight to the anthropocentric and individual-centred worldview that regards all non-human creation from an instrumental and profit-focussed perspective, that is to say, a worldview that sees everything as separate entities in a game of profit whose primary purpose is to satisfy human acquisitiveness and other desires.

On this basis, Nasr argues that humanity does not own nature but has received it on loan. Any intervention or damage to nature must, according to this view, happen with the permission of the owner/God. Various Qur'anic verses are invoked to underscore the point that humanity's mission is to administer and take care of nature and its resources in the best way possible (Qurʾan, 57:7). Custodial responsibility and humanity's role as God's servant are presented as a combination that jointly offer a guarantee that humanity will respect and protect non-human nature from plunder.

The position of power that Islam's creation theology accords to humanity is conditional upon that custodianship taking place in accordance with God's law. Humanity's role as custodian does not give it the right to interfere with nature without the owner's permission. There are no circumstances under which humanity can damage, destroy, pollute or in some other way harm God's creation. In order to gain the owner's, that is to say, God's, permission to exploit the resources of earth and sea, humanity must follow God's commandments and avoid those things that God has forbidden. Those who wish to obey Islam's commandments and avoid its prohibitions must follow Islamic law, *Shariʿa* (Javadi Amoli, 2007).

3.5 Muslims and the Embracing of Western Ideals

A central concept in Nasr's thinking is that although Muslims have been severely affected by environmental destruction and climate change, they bear no blame for environmental destruction and global heating. Environmental and climate problems in Muslim countries are the result in part of global climate conditions but to a greater degree because of the fact that Muslims have embraced the Western ideal of modernization, its consumption patterns and lifestyle, and thereby themselves become a part of the problem.

According to Nasr, an important reason for this is that in Muslim countries since the eighteenth century the Muslim worldview has been driven to the periphery. This took place at the same time as a number of Muslim countries found themselves under European colonial rule and absorbed the European civilization's economic (read: capitalist) ideals. Many Muslims have adopted the Western view, which rests on the idea that humanity should rule over nature and exploit its resources for its own economic benefit (Nasr, 1993). In light of this, Nasr draw the conclusion that for the last few centuries Muslim countries have not been following the Muslim worldview with regard to its ethical norms and knowledge ideals.

According to Nasr, the reactions of the Muslim world can be divided into two opposing voices. One is the Salafist/Wahhabi voice whose main characteristic is opposition to certain aspects of Western civilization in the areas of political and individual rights and freedoms as well as of social and cultural reforms. They have, by contrast, adopted the Western lifestyle and its consumption patterns. This resistance has, however, proceeded from a legalistic viewpoint. Within this intellectual tradition, the most important factor is the conservative and literalist interpretation of Sharia law and its representatives in all aspects. This approach is very far from what Nasr defines as classic Islamic theology and philosophy and its rationalistic theological perspective. He accuses Salafists/Wahhabis of being incapable of discovering and grasping what is harmful in the Western worldview. This inability is also the reason why Wahhabism's leading exponent, Saudi Arabia, and the so-called Gulf states are today the greatest consumers of Western products. When they mastered the art of managing the legalistic obstacles, they were free to embrace Western consumption patterns without any hindrance (Nasr, 1993).

The other voice belongs to Muslim modernists, who celebrate everything that is associated with the West. Their prescription for success is to copy and incorporate into Muslim countries everything that has happened in the West. The colonization of Muslim countries and the collapse of the Ottoman Empire sent shockwaves through Muslim countries, presenting Muslim thinkers with an array of new challenges. The prevailing Muslim paradigm had also been shaken to its core (Jafarian, 2015). It was at this moment that Muslim thinkers presented their various proposals for how to respond to these challenges and how to solve the crises that had afflicted Muslim countries. Some wanted to turn the clock back to the earliest era of Islam; others wanted to reinterpret the sources of Islam in new ways. Those that came to be known as modernists chose to combine these two models.

Even though they were highly critical of the European colonial powers, they directed their sharpest criticism to their own Muslim leaders and the religious authorities. They accused them of having betrayed Muslims and of having deviated from true Islam. Their principal message was that people should adopt Western science and its industrial progress in order to create prosperity in Muslim countries. The industrial revolution and Western technology were applauded by Muslim modernists who did not question their harmful impact on the environment (Nasr, 1993).

These opposing reactions served to marginalize the original Muslim worldview whose principal hallmark is a love of nature and a desire to decipher its mysteries. Another area that was marginalized was Islamic ethics, whose normative criteria for moderation in consumption and all other areas was displaced by a Western lifestyle promoting extravagance and overconsumption.

Nasr nonetheless argues that Islam's religious worldview cannot be reduced to a moral voice. It presents a rational voice whose primary characteristic is that it levels severe criticism at anthropocentrism and the Western science that regards itself as having a monopoly on rational thought and that has been a driving force behind controlling and wielding power over nature (Nasr, 1993).

Another point raised by Nasr is that Muslim countries were forced to accept the Western model in order to address their economic problems. The colonial powers and their successors imposed the Western lifestyle upon peoples around the world, including the Muslim world. It is a wasteful lifestyle that devours vast quantities of energy and raw materials. Muslims' capitulation to the West meant that the Muslim theistic view of nature was suppressed in favour of anthropocentrism, individualism, and a highly destructive attitude towards nature and its resources (Nasr, 1993). Indeed, Nasr's line of argument is that the Muslim theistic worldview, its educational ideals and accompanying institutions underwent a kind of collapse. As he sees it, this distancing is having a profoundly corrosive impact on the younger generation, who are becoming more alienated from Islam.

The situation is not hopeless, however, according to Nasr. Despite the European and Western dominance over Muslim countries, there is still hope for Muslim countries, he emphasizes. Unlike secularized Europe and the West, where the religious worldview has entirely capitulated to science, the situation is different in Muslim countries. The Muslim world has all that it needs in order to deal with these problems. The religion of Islam still has great credibility with the general public. Natural theology and its view of the sacred in a Muslim context has not suffered the same setbacks as has Christianity in the West. Islamic ethics based on revelation have a strong standing among Muslims. These circumstances make it easier to engage Muslims in the work of saving nature and the environment from all of the harm that has come from the West, argues Nasr (1993).

In order to succeed in this endeavour, Islam's view of nature needs to be presented in a way that is accessible to the larger Muslim public. This is necessary in order for Muslims to be able to critically assess the West's excessive faith in science and its scientistic reductionism. The Muslim public also needs to gain insights into Islam's scientific history in order to who that there is no lack of a tradition of scientific knowledge in Islam (Nasr, 1993).

Another step in this direction is to disseminate knowledge about Islamic Sharia' ethical handling of questions relating to nature and the environment. This includes greater knowledge of Sharia law's prescriptions relating to the protection of nature, ethically defensible animal husbandry, and the ethical handling of plant life, natural resources, etc. According to Nasr, Islam offers a worldview that involves peaceful coexistence between humanity and God's non-human creations: nature, animals, and plant life. Islam advocates a loving relation to nature. This worldview forms the basis for an ethical approach which originates in Islamic values that prevent humanity's unrestrained exploitation of nature and the environment (Nasr, 1993).

In order to counteract the uncritical embrace of the Western worldview, Nasr argues, the Muslim public must be made to realize that nature does not consist solely of God's signs but also reflects the divine. This view he is based on the following verse:

> And God's is the east and the west; therefore wherever you turn, there is the face of God (Qur'an 2:115).

He interprets this verse as meaning that everything in the world is connected and that the destruction of a part of God's creation will affect other parts of the world. Humanity must realize that it cannot save nature without recognizing the divine in nature. A first step towards gaining this insight into the divine in nature is the insight of the sacred in creation. Humanity cannot gain recognize the sacred in creation without discovering the sacred within itself and the sacred as such. These insights are important preconditions if humanity is to live in harmony with nature. According to Nasr, it is not possible to manage the threat to nature, the climate, and the environment without curing the spiritual crisis of modern humanity itself. This, he argues, can only happy if humanity rediscovers the sacred in both itself and nature (Nasr, 1993).

3.6 Conclusion

A central theme in Islamic ecotheology, which in this chapter has been illustrated primarily through the ideas of the philosopher Nasr, is its sharp critique of Western civilization. This finds expression in its critique of the anthropocentric worldview, Cartesian deism, what it identifies as excessive faith in science, and the libertarian philosophical and political perspective. These are presented as significant factors that have led to a desacralizing of the cosmos, an unbalanced focus on the interests of humanity that promote humanity's individual autonomy and freedom of choice. This is a perspective that builds on a kind of human exceptionalism: a secular view that has erased all that is sacred in the world.

It is important to note in this context that Nasr is not only advocating an Islamic ecotheological perspective. His thinking is equally influenced by the religious-philosophical intellectual tradition of *philosophia prennis*, which views religion in general as the main avenue for insight into God or the Supreme Being. This intellectual tradition holds up metaphysics as a mode of knowledge with which humanity can interpret the cosmos. Much of Nasr's critique of science and the anthropocentric perspective has its basis in this religious-philosophic tradition.

Nasr is not alone in this opinion, however. Criticism of this perspective also come from other quarters. It is directed at what is described as humanity's self-sufficiency, which proceeds from the idea of humanity's independence from God. It is an opinion that, according to the British Muslim philosopher Shabir Akhtar (1960–2023), encounters resistance in Islam's natural theology.

> The Qur'an attacks human presumption, such as claims of human self-sufficiency and hubris and creaturely independence from God. It sets limits to the use of human reason in the aftermath of revelation, after the closing of the final canon. It *reveals* to us the correct uses and limits of unaided human reason, the correct contours of a natural theology, and the general apologetics that would help defend revealed claims about God against the secular detractors of Islam (Akhtar, 2020).

There is a very solid suspicion here of excessive faith in science. Without denying the importance of science and technology, what is described as scientism's reductionist perspective has been the object of sharp criticism. Nasr characterized the latter as a totalitarian ideology that does not recognize other perspectives and views.

Nasr questions the Western model's claim to universalism. He emphasizes that Western science, technology, and philosophy have gained universal status through colonialism and modern globalization. Other contributory factors behind this development, it is argued, have been the implementation of Western political ideals, the popularizing of its scientific advances, and the social changes that have taken place in the wake of globalization. Globalization has also affected culture, language, communication, norms and values, social relations, and everyday practices in large parts of the world, including the Muslim world generally. This is a development that Muslim countries have been unable to resist.

In summary, these factors have been identified as factors that have contributed to the environmental and climate problems that have afflicted Muslim countries. The heart of the problem is seen as lying in resistance to religion, spirituality, and the abandoning of traditional norms and principles as embodied in the authentic form of Islam (Nasr, 2001).

Islamic creation theology is offered as a non-anthropocentric substitute and an alternative approach to Western anthropocentrism and what critics characterize as its dogmatic faith in reason and the scientific perspective. Nasr articulates a critique that wishes to settle accounts with the desacralized worldview and put a stop to the secular worldview's monopoly, which, he argues, accords monopolistic status to science (Nasr, 2007). This will happen through a return to a theistic worldview that, by strengthening humanity's faith and resacralizing the cosmos, will limit the climate threat and environmental destruction. This resacralization is only possible within the framework of a worldview in which the theistic worldview has a central place. The idea is that humanity in a post-secular era, through faith and recognition of the divine in both humanity and God's non-human creation, will arrive at the insight that it must manage the world's resources in a moderate and prudent fashion. It is an alternative that protects both the human and the non-human.

Islamic ethics is held up as an important means for revitalizing the sacral worldview. Advocates argue that it brings a spiritual dimension to the relationship between humanity and the cosmos. This is because Islamic ethics urges humanity to pursue moderation and reminds it to recognize its role with regard to God's non-human creation. Even more important in this respect are the foundational concepts in Islamic creation theology: custodianship, *Amana*; testimony, *Shahada*; humanity's role as God's servant, *ʿAbd*, and God's representative on Earth, *Khalifa*; and the view of both humanity and nature as God's signs, *Āyāt* (Nasr, 1993, p. 134).

Nasr offers these principles of Islamic creation theology, in combination with Islamic theology, as a guarantee that humanity should not engage in activities that are destructive for the environment and climate. This approach represents a sustainable model that acknowledges non-human creation and show that humanity's fate is closely bound up with that of the non-human. On this basis, Nasr rejects the accusations levelled at Islam's creation theology, claiming that it regards humanity's role as an administrator of God's creation, which has been lent to humanity for its earthly existence. This, he argues, is a perspective that leaves no possibility for the plundering of nature. Moreover, humanity must use the resources of nature, that is to say, God's creation, in accordance with the permission of the owner, God, which means that God the creator has not given permission for activities that have a negative impact on creation (Farahanifar & Farahanifar, 2013).

Another aspect of this debate is that Nasr's view on the notion of custodianship and similar ideas as unique to Islamic creation theology. In fact, many of these ideas are to be found in the work of Christian theologians, among others. This shows that the dichotomic relationship between Islamic ecotheology and other theological traditions, as Nasr and other Muslim thinkers present it, does not match the current ecotheological debate over environmental and climate threats. Many of the ideas that they portray as unique to Muslim ecotheology have their equivalents in other religious traditions. Moreover, it should be noted, that scientific researchers are working hard to minimize these threats.

Another problem relates to the image of the divine that Nasr articulates. A number of feminist Christian theologians have criticized the image of the divine as it figures in traditional Christian theology and in Nasr's writings, specifically the image of the divine that feminist Christian theologian Sallie McFague (1933–2019) describes as the "monarchist model of God". This is an image of the divine as something wholly different in kind from the world, elevated above it, and its absolute ruler. The fundamental characteristic of this image of the divine is the view of God as a ruler who is either dominant or benevolent, or both (McFague, 1993).

This traditional image of the divine has long been questioned by thinkers engaged in the feminist theological debate in Christianity. In particular, critique has been directed at its implications for how we view humanity and the kind of political order or orders that it legitimates. This is an image of the divine that presupposes that God is supreme and that everything else is insignificant. McFague argues for another image of the divine, which has abandoned the archaic and monarchic image of the divine. Instead of the dichotomic image of ruler and world ruled by God, she proposes that we think of the world as "God's body," in which an "intimate" relationship exists between God and the world (McFague, 1993).

This latter image of the divine is more compatible with the image of the divine that we encounter in, for example, the Sufi intellection tradition within Islam, which, taking its point of departure from Islam creation theology's conception of humanity as having a dual nature, sees a clear connection and palpable affinity between God and humanity, in which humanity has an affinity with God by virtue of its spiritual side. It is on this basis that Sufists speak of humanity's absorption or reunification with God.

Another problem with Nasr and other Muslim ecotheologians is their view of environmental problems in Muslim countries. These thinkers are united in their unwillingness to admit that Muslims bear any responsibility. Responsibility is exclusively laid at the door of the former colonial powers, globalizing modernity, and Muslims' embracing of Western lifestyles. This approach, for all its criticism of the Western worldview and the former colonial powers, lacks a self-critical perspective of its own. Rather, it prefers to present Muslims as victims of the Western worldview's claim to universalism and the globalization of its lifestyle.

It is a characteristic of these thinkers that they seem to take for granted that discussion of the positive aspects of the Muslim perspective will by itself result in positive measures. The Muslim perspective is presented as an ideal, and when they are confronted by the fact of environmental and climate problems in Muslim countries, they entirely exonerate this idealized Muslim perspective of all responsibility. The absence of this good/ideal state is not seen as having anything to do with the ideas themselves. Blame is instead laid at the door of people/Muslims who have not understood or managed to implement this ideal perspective, which derives from the authentic form of Islam. This line of reasoning gives the impression of being utopian and lacking any basis in reality.

This uncritical self-image in turn serves to explain why Muslims have embraced a globalization process that is centred on the West. One explanation starts from a defensive position that portrays Muslims as victims who have been unwillingly drawn into this. For this reason, they cannot be expected to bear any responsibility: the environmental and climate crises can only be solved and managed by those who caused them in the first place.

The other explanation takes its point of departure in the utopian notion that reproaches Muslims for having distanced themselves from the authentic model of Islam. This explanation is also focused on structural and political factors. Muslim authorities and those in power bear the blame for having failed in their responsibility to implement the ideal Muslim model. Both explanatory models make no effort to problematize the issue of individual responsibility. The problem with focussing on structural, political, and collective measures is that this perspective hamstrings those measures that might be pursued on an individual level, including the many measures, perspectives, and lifestyles that engage individuals beyond the Muslim world that might be adopted by individual Muslims. The two levels do not exclude each other. Both are important for the task of responding to the environmental and climate threats that endanger the future of humanity.

References

Akhtar, S. (2020). Can an Islamic natural theology explain God's silence today? *Renovatio, The Journal of Zaytuna College, 4*. https://renovatio.zaytuna.edu/article/can-an-islamic-natural-theology-explain-gods-silence-today

Farahanifar, S., & Farahanifar, M. A. (2013). Moḥiṭ-e zīst va nezām-e akhlāqī-ye ān dar islam (The environment and Islamic ethics). *Maʿrifat-e Eghtesādī-e Eslāmī, 4*(2).

Jafarian, R. (2015). Salafigari zendeh ast. Gooneh shenasi-ye andishe-ye bazgasht be Qor'an dar tarikh-e andishe-ye eslam va Iran. *Mehrnameh, 44*, 189–191.

Javadi Amoli, A. (2007). *Eslam va mohit-e zist* (Islam and the Environment). Asra Qom.

McFague, S. (1993). *The body of god: An ecological theology*. Fortress Press.

Nasr, S. H. (1993). *The need for a sacred science*. State University of New York Press.

Nasr, S. H. (1996). *Religion and the order of nature*. Oxford University Press.

Nasr, S. H. (2001). Hela naturen talar om Gud. In *Världsreligionernas syn på naturen och på miljökrisen som spegling av människans inre kris*, Swedish translation by Björn Sahlin, Stiftelsen Abrahams barn.

Nasr, S. H. (2007). In the beginning was consciousness. In S. H. Nasr & W. Chettick (Eds.), *The essential*. World wisdom Inc..

Mohammad Fazlhashemi is a Professor of Islamic theology and philosophy at Uppsala University, Sweden. Fazlhashemi's research interests concern theological and philosophical ideas, political ideas with links to Islamic theology, philosophy and legal/judicial interpretations. He has also studied the perception and the construction of the images of Europe/the West and modernity among Muslim thinkers.

Chapter 4
Re-thinking Colonialism, Globalization, and the Displacement of Wild Thinking: Postcolonial and Indigenous Perspectives on Religious Education and Sustainability

Kathrin Winkler

4.1 Introduction: Sustainability and Education

In September 2015, at the United Nations Sustainable Development Summit, Member States formally adopted the 2030 Agenda for Sustainable Development in New York. The agenda contains 17 goals including a new global education goal (SDG 4). SDG 4 is 'to ensure inclusive and equitable quality education and promote lifelong learning opportunities for all' and aims to empower all people to help shape a sustainable future by 2030. The formal education settings of school, vocational training and higher education have a central role to play, as they shape much of the everyday life and later development of children and young adults. In the midst of current multiple sustainability crises, such as climate change, species loss, and growing social injustice, formal education impacts learners' knowledge, attitudes, emotions, and values, and can thus help them to build the future.

Grund and Brock (2022) show that there is a mismatch between scientific findings and recommendations for sustainability (cf. Armstrong McKay et al., 2022; Steffen et al., 2015) and the reaction of formal educational structures to this scientific knowledge. The content and structures of the education system are often designed to continue the established, and to provide continuity. Formal education thus proves particularly unwieldy in the face of deep, necessary change in times of upheaval (Gräsel, 2010). This can be understood by the fact that sustainability has so far not been very visible in formal education and has mainly been taught by individual committed teachers and in thematically relevant subjects such as chemistry, biology, or geography (Grund & Brock, 2018). The basic ideas and structures for

K. Winkler (✉)
Religious Education and Religious Studies, Lutheran University of Nuremberg, Nuremberg, Germany
e-mail: kathrin.winkler@evhn.de

S. Windsor, O. Franck (eds.), *Intersections of Religion, Education, and a Sustainable World*, https://doi.org/10.1007/978-3-031-81809-7_4

contemporary educational goals, through which students can participate in future social progress, still seem largely untouched by the necessary changes. Looking at the goals defined in the various educational sectors (e.g. in school laws or in the qualification goals for teachers), areas such as the protection of livelihoods and the promotion of social justice have hardly been strengthened in recent years, despite the current climate changes (Brock & Holst, 2022; Holst, 2022; Holst & Singer-Brodowski, 2022).

This stagnation is also reflected in current educational practice, where a predominant pessimism about the future is evident among learners and teachers due to the lack of effective opportunities for action and a conceptual understanding of sustainability in the education system. Often the size of the problem seems overwhelming compared to the perceived very manageable scope of one's own actions. Currently, negative emotions related to the sustainability crises, such as worry, frustration and fear are already widespread among the younger generation (cf. BMUV & UBA, 2022) and are described, among others, by the term solastalgia (cf. Albrecht, 2019).

Referring to the basic orientation of formal educational institutions, Orr (1994) has already aptly described that contemporary education ultimately empowers learners to become "more effective vandals of the earth" (p. 5). This analysis draws attention to the extent to which formal educational institutions, through their teaching of values and preparation for participation in resource- and consumer-based social success, are also part of the problem, sometimes implicitly and unintentionally, rather than just part of the solution. In this context, against the backdrop of socio-ecological threats, UNESCO's Assistant Director-General for Education notes the increasingly pressing question of whether what people learn is "really relevant to their lives" (UNESCO & Deutsche UNESCO Kommission (DUK), 2021, p. 2). Thus, Education for Sustainable Development aims to provide everyone with the attitudes, competencies, knowledge, and values needed for a livable future (Goodman et al., 2022). Because ESD aims at broad empowerment, namely of everyone, there is a fundamental requirement to realign underlying goals, values, and ultimately "all levels of education and learning to contribute to sustainable development" (UNESCO & Deutsche UNESCO Kommission (DUK), 2021, p. 12). However, studies show that while learners have a keen interest in the topic of sustainability and anticipate negative impacts on their own lives, at the same time, the majority of learners do not experience school, college, or training as empowering them to effectively shape a sustainable future.

Exactly at this point the central question arises: what significant role can religious education play in the context of education for sustainable development and which ethical-theological, but also political implications, consequences and educational practices can result from this? In particular, this includes identifying a deeper understanding of the connection between religious educational traditions and the responsibility to live sustainably, which can contribute to global learning in the sense of a good life for the whole globe. Consequently, this also means for religious education to include postcolonial analyses that reveal that local educational systems and indigenous knowledge were delegitimized and consciously suppressed in the course of European colonization and missionization. This relates specifically to the traditional knowledge of indigenous communities for a healthy and resilient society and a sustainable way of life, which was deleted during European colonization and the associated normative concepts of education (Lephakga, 2015; Chimee, 2018). Making these disqualified bodies of knowledge audible and accessible again and bringing them into dialogue with one's own religious affiliation is the goal and means the following: Empower learners to act from the value-based sources of religious traditions, deal constructively with increasingly common sustainability-related emotions, such as fear, sadness, and hopelessness (cf. Grund & Brock, 2022), and develop an understanding of sustainability and sustainable education. The contribution of religious education to sustainable development education is therefore explored in a three-step approach.

In a first section, the connection between colonialism, the missionary movement, globalization, and the destruction of the ecosystem, which led to the extinction of indigenous knowledge, is investigated. Subsequent to this, the term solastalgia is introduced, referring to the enormous emotional consequences and pain about the destruction of nature and the exploitation of natural resources (Sect. 4.2). The significance and the practice of hope are then opened up to discuss how indigenous epistemologies and spiritualities can help to shape a new story of interdependent connectedness and live this new story through practices of embodied hope (Sect. 4.3). Finally, a conceptualization of what this process of insight can contribute for religious education and sustainable development is given (Sect. 4.4).

4.2 What Has Been? – Historical Developments

The first step involves identifying historical and theological developments that have contributed to the ecological crisis and now make re-framing urgent. On the one hand, this includes, addressing the entanglements of Christian missionary societies in the *colonial domination structures of Europe* that helped to legitimize the exploitation of natural resources. In the name of Christianity, the suppression of indigenous bodies of knowledge and knowing, which contain a treasure of care for nature and forms of sustainable living, have been legitimized (World Council of Churches, 2012). It is about the realization that the ecological crisis is a product of colonization.

4.2.1 Colonialism, Globalization, and the Displacement of Wild Thinking

The potential to restore the unity of life and achieve true sustainability must be a scientific, ethical, cultural, theological, and decisively an educational response. That this is more necessary than ever, but not a matter of course, is summed up by the

Australian philosopher Clive Hamilton: "Today the greatest tragedy is the absence of a sense of the tragedy" (Hamilton, 2017, p. x). The challenge of transforming these human but also institutional mindsets and generating the insight that the globalized high-growth society cannot be continued in this way is also confronting the education system and, with it, religious education. If religious education wants to convey its social relevance and initiate change processes in the ecological crisis from the sources of religions, it requires strenuous processes of conversion and redirection. From a Christian perspective, this implies above all a critical deconstruction of historical developments of Christianity and practicing theological self-criticism until a new theology leads to practical consequences, in particular in educational processes.

This includes first taking a scrutinizing look at the colonial and missionary societal entanglements of the Christian Western world (cf. Aly, 2021; Döbler, 2021), which played a crucial role in making the current destruction of the ecosystem possible at all. Uncovering the traces of colonial structures of dominance and bringing to light marginalized perspectives in today's societies, but also making visible structures of neocolonial globalization that are co-responsible for the exploitation of natural resources, seems more than ever also an urgent theological task (Gruber, 2018). Postcolonial theologies call to mind that the Bible is the central reference point of the Christian and "civilizing" mission that accompanied and first legitimized colonial hegemony. The quote attributed to Anglican Archbishop and Nobel Peace Prize laureate Desmond Mpilo Tutu clarifies this relationship: "When the first missionaries came to Africa, they had the Bible, and we had the land. They asked us to pray. And we closed our eyes. When we opened them again, the situation was just the opposite: we had the Bible, and they had the land." Examining and uncovering this connection between knowledge and power in the textual production of the West, therefore, emerges as a necessary approach to theological reflection (Sugirtharajah, 2013, p. 125).

At this point, no differentiated historical insight into the entanglement of mission and colonization can be given (cf. Pittl, 2018; Ratschiller & Wetjen, 2018; Habermas & Hölzl, 2014). Nevertheless, in historical retrospect it is evident that in the course of colonization and Christianization, the spiritual knowledge and worldviews of indigenous peoples were judged to be primitive, pagan, and anti-Christian. In many cases, the Christian mission provided the theoretical superstructure for this by conveying that the religious as well as societal superiority of white people was the God-ordained order and that indigenous cultures were not only inferior but also sinful and devilish. This hermeneutic practice laid the foundation for suppressing indigenous knowledge of animate nature, enslaving their mediators, understanding them as a lower race, and speaking of white civilized superiority (World Council of Churches, 2012). European colonization and Christian missions, which have proceeded against indigenous worldviews since the Middle Ages, can be understood as a crusade against indigenous epistemology through Christian theology of redemption, which removed humans from the context of nature (Kirchhoff, 1999). Until the twentieth century, supported by ethnology and the scientific traditions of rational reason, the binary structures of the Eurocentric worldview were maintained: here the civilized rational Christian world, there the archaic primitive pagan world; here the culture, there the savage people who had to be developed. It is these interpretations of the world that continue to form the foundations for racism, exclusion and, above all, the exploitation of natural resources.

French ethnologist Claude Lévi-Strauss (1968), who was one of the first in the twentieth century to address the displacement of indigenous thinking forms, initiated a phase of deconstruction of colonial structures of power and thought. With his re-reading of indigenous narratives and forms of thought, he provided an impressive insight into the holistic, often symbolic, worldview of indigenous cultures, which he understands as an equal approach to reality alongside the scientific findings of modernity. In his ethnological studies in Brazil, Canada, and Japan, he comes to an important conclusion: What Lévi-Strauss calls "wild thinking" consists of important forms of thinking that are not inferior to those of science and that do not need to be disciplined according to the standards of rationality. Wild thinking "forms a precisely articulated system and is in this respect independent of the other system that science will later establish." (Lévi-Strauss, 1968, p. 25). For Lévi-Strauss, these two forms of thinking do not form an opposition, but rather show themselves as two ways of knowing from a different point of view. From an ethnopsychological point of view, indigenous knowledge is linked to timeless mythological concepts: the material world is alive, a soul substance is seen in it, which weaves through everything, and which is experienced as a kind of cosmic religiosity. Humans are woven into this network of life and mandated with responsibility (Lévi-Strauss, 1968). This thinking incorporates an approach to the world and a sense of meaning that have enabled indigenous cultures to live sustainably for thousands of years. Indigenous knowledge and practice are more necessary today than ever because the way of life and the millennia of care and responsibility for nature can be an important key to the survival of humanity. Bringing Indigenous knowledge and wisdom into dialogue with Christian theology can become a healing re-reading of societal as well as theological power traditions that also deconstructs neo-colonial efforts, e.g. in the form of 'green colonialism', which implements its renewable energy needs and other conservation activities on the territory and at the expense of indigenous groups.

4.2.2 Solastalgia and the Loss of Solace

The preceding reflections not only indicate the societal and global consequences, but also draw attention to profound emotional consequences of the ecological crises. It is the existential sense of loss that people and in particular young people all over the world express as emotional pain about the destruction of nature and the exploitation of natural resources (Ellis & Cunsolo, 2018). This pain is exemplified by the school strike for the climate, which started as a global social movement from students and led to the founding of Fridays for Future. The initiator Greta Thunberg showed this emotional pain at the UN Climate Summit in 2019 when she stated: "People suffer, people die, whole ecosystems collapse. We are at the beginning of a mass extinction and all you can talk about is money and the fairy tales of economic growth that will last forever. (...) How dare you steal my dreams and my childhood with your empty words?"

This sorrow and emotionality was summed up by Australian environmental philosopher Glenn Albrecht as early as 2003 in the term solastalgia, which he coined anew in light of the emotional reactions of people experiencing intensive coal mining in Australia's Hunter Valley in the early 2000s. As the mines expanded, Albrecht registered a basic tenor in the emotional reactions of the valley's inhabitants: they knew that the mines were the cause of their suffering, but they found it difficult to put the feelings into words. The physical disturbance of the valley corroded the sense of home that people had felt. And while the mines dyed more and more green fields gray, Albrecht gave the sense of loss a name: "Solastalgia" denotes the pain of losing a comforting sense of home safety. The neologism is composed of the Latin solacium ("consolation") and the ancient Greek algos ("pain"). With this neologism, Albrecht describes a form of physical or existential stress, which is caused by environmental changes and especially environmental devastation. "Solastalgia is the lived experience of the loss of value of the present and is manifest in a feeling of dislocation, of being undermined by forces that destroy the potential for solace to be derived from the immediate and given. In brief, solastalgia is a form of homesickness one experiences when one is still at home." (Albrecht, 2006). This existential distress is experienced in different ways across the globe and is already precipitating the loss of home, identity and culture through climate induced displacement forcing movements of people in search of new homes: Pacific islands that are sinking, rainforests that are being clear-cut for soybean farming and cattle ranching, droughts that massively affect agriculture, the retreat of permafrost in the Arctic that is displacing indigenous people and encouraging new economic exploitation on their land, etc. "Let's get it out of the world. Let's eliminate the circumstances, the forces that cause solastalgia." Albrecht posits that the antidote to solastalgia is "in a future that has to be designed and created" (Albrecht et al., 2007, p. 45). In this way, the Australian environmental philosopher provides the necessary impetus to develop counter-designs that can have a decisive impact in (religious) educational processes. What role or insights might hope offer in such processes in light of the ecological crisis?

4.3 What We Can See Now? – The Crisis of Hope

The engagement with these historical lines of development, however, opens up the opportunity to re-engage indigenous narratives of sustainability in religious education processes and to make their alternative world relations constructive, which is fed by a dialogue with indigenous epistemology and spirituality (Lam et al., 2020). Indigenous communities have long been raising their voices to share their knowledge in the ongoing fight against climate change and the growth society. By means of their local practices and initiatives, they share in educational projects the importance they attribute to cultural values such as respect for nature, sharing in a community, and local traditional knowledge for a healthy and resilient society and a sustainable way of life (cf. Indigene Wege, 2020).

4.3.1 Embodying Hope in Religious Educational Practices

In this time of extraordinary ecological crisis alongside of which there is increasing awareness of solastalgia or eco-anxiety, as Timothy Robinson observed in a recent article, there is a "manifesting in a crisis of hope" (Robinson, 2020, p. 1). The problem with hope is, as Ruth Guyatt states, its "ambivalent duality" (Guyatt, 2020). Hope is associated with and understood to be passive optimism, or blind faith which manifests in avoidance, denial, complacency, and inaction (Anderson, 2006; Guyatt, 2020; Head, 2016; Robinson, 2020) particularly in response to climate change and the unfolding ecological crisis. According to Robinson (2020) the accounts of hope in "recent environmental literature and in traditional Christian formulations rely on faith in political will, technological innovation, or an omnipotent divine sovereign to intervene and save" (Robinson, 2020, p. 1). But Robinson contends "such accounts are inadequate for this moment" (Robinson, 2020, p. 1). A helpful study on *How hope and doubt affect climate change mobilization*, by Marlon and colleagues reveal two taxonomies of hope: constructive and false hope (Marlon et al., 2019). People who rely on false hope believe God or nature will solve the problem without the need for human intervention. In contrast, constructive hope arises from the belief that humans are capable of changing their behaviour (Ortiz, 2020).

Maria Ojala provides further clarity on constructive hope explaining that "constructive hope arises when a person understands the seriousness of climate change and concurrently feels there are positive goals within reach for addressing the issue" (Ojala, 2012, p. 627). Ojala was particularly interested in whether a sense of hope among young people is positively related to pro-environmental behaviors, or whether it is simply a sign of illusory optimism (Ojala, 2012). Her research findings showed that young people who have constructive hope about climate change are significantly more likely to engage in positive environmental behaviors, compared with young people who lack hope or whose hope is based on denial of climate change. Ojala identified three main sources of constructive hope. The first is the ability to describe their concerns about climate change and to reframe them in a different and positive way. The second is identifying trust in sources outside oneself, such as technology or environmental organizations. The third source of hope is the trust in one's ability to make a difference. Ojala concludes that "hope about climate change is more than an illusion or solace; it could also be an important path to engagement" (Ojala, 2012, p. 638). Both studies demonstrate the critical importance of the source, habitus, and place of hope. For religious education, they set the course for how to focus on hopeful agency and pedagogically address the ecological crisis with children and youth (Winkler & Carroll, 2021).

Specifically, that implies rethinking hope which begins with the "relinquishment of false optimism" (Robinson, 2020, p. 1) and the acceptance of the reality and enormity of the ecological crisis that confronts us. As Chris Doran has so aptly captured, in embracing hope one needs also to cultivate the ability to "know how to talk about hopelessness" (Doran, 2017, p. 9). Robinson asserts any "credible account of hope will begin with truth-telling, which includes accepting what climate scientists are telling us about the future unclouded by false hopes and coming to terms with human responsibility for our current condition" (Robinson, 2020, p. 7). Borrowing from Miguel De La Torre (Torre, 2017), Robinson argues that the key to relinquishing false optimism is to embrace hopelessness and the feelings of grief, despair and helplessness that accompany it. And vice versa this means that hope is not exclusive of feelings of despair, hopelessness, or grief (Franzen, 2019).

Hope consists of a broad range of emotions including painful ones (Head, 2016, p. 74). Lesley Head, drawing on Anderson's geography of hope, develops an understanding of hope premised on the following four aspects. She understands hope as, firstly, a process that creates possibility and potential, or at least opens up spaces in present day reality for things to be done differently. Secondly, hope is not exclusive of melancholy and grief. Thirdly, hope risks disappointments and has no guarantees. Finally, hope is grounded in the everyday and not in some distant future. All these four aspects are embodied or find expression in practices. Head contends that hope is something to be practised rather than felt. The basis for hope, therefore, is in practices rather than emotions. In this way, hope becomes active, engaging and fully present in the world. Brian Treanor articulates this well in stating "to hope is to adopt a certain way of being-in-the world; it is an active, not passive disposition" (Treanor, 2018, p. 11). Hope, "requires our active participation in both the enjoyment of the world and the maintenance and promotion of its goodness" (Treanor, 2018, p. 22). This posture provides a basis for how we can imagine and live in the world differently. Developing and implementing these seems to be a central pedagogical task of religious education in the present day (Winkler & Carroll, 2021).

Embodied practices of hope in the Anthropocene (Bertelmann & Heidel, 2018) will require not only, but in particular also from religious education a new story or narrative that values our *interdependent interconnectedness* with the whole of life (the human and non-human creation). The old story powered by colonialism, capitalism justified by Christian narratives of dominion, domination and power have not served humankind well and has led to the degradation of both the environment and the fabric of human relationships. The future and history are written and shaped by the imaginations of those in the present fuelled by their values and vision of what the world could be in the future. In this context, religious education processes can be crucial in creating intentional spa ces and platforms for such new imaginings of the future to be imagined and realized (Winkler & Carroll, 2021).

In conclusion, hope is cultivated and sustained by the recognition and awareness that we are part of the larger whole grounded in an expanded notion of the Sacred. The resources for engaging hopefully with and in the climate crisis do not and cannot rest in one religious tradition alone. A new and robust account of hope requires the wisdom of "religious and non-religious accounts of spirituality that find the sacred embedded within the world..."(Robinson, 2020, p. 8). It is here that indigenous epistemologies and spiritualities provide a lens through which to live hopefully in the present in view of the future (Winkler & Carroll, 2021).

4.3.2 *Embodied Practices of Hope in Pacific Indigenous and Faith Based Re-storying*

Indigenous epistemologies and spiritualities play an imperative role in addressing the climate crisis and the loss of biodiversity and as such the inclusion of indigenous perspectives in religious education processes is critical. Indigenous communities occupy 20–25% of the Earth's land surface, and 80% of that land mass holds the world's remaining biodiversity (Raygorodetsky, 2017, 2018). These figures show that Indigenous Peoples of the world have been and still are the protectors of ecosystems. They have been caretakers of the Earth for centuries, but their knowledge has either been devalued or ignored. The Special Rapporteur on the rights of Indigenous Peoples, Mr. José Francisco CalÍ Tzay asserts "indigenous peoples and their traditional practices, are key to achieve sustainable development, combat climate change and the conservation of biodiversity"(UNHCR, 2020). Indigenous knowledges and perspectives are crucial not only for addressing the burgeoning climate emergency but also for navigating the way forward to a hopeful post-COVID, post-growth and post-fossil fuel future.

From this it becomes evident that indigenous epistemologies and spiritualities can help humankind to frame and shape a new story of interdependent interconnectedness and to live this new story through practices of embodied hope. Indigenous communities understand and remind us of the fragile yet vital relationship between human beings and the environment and the critical importance of keeping this relationship in balance (Vaai, 2019; Vaai & Casimira, 2017). Although diverse and varied in traditions, cultures and practices, indigenous peoples hold and value the sacredness of our interdependent interconnectedness with the whole of life. They understand epistemology, spirituality and being as an integrated whole inclusive of the environment (Bird et al., 2020; Meyer, 2013) and that the health and well-being of the earth and human beings are intimately interwoven. Reconnecting the whole of life necessitates a new narrative that shapes how we may live differently into a new future. A case in point is the following example from the Pacific where embodied hope is practised through re-storying the household from a Pacific perspective in anticipation that a new and different future may be realized (Winkler & Carroll, 2021).

The Pacific, 'the liquid continent', has been identified as one of the world's climate change hot spots (Boege, 2018, p. 2). The "most significant effects of climate change include reductions in agricultural productivity; reductions in water quantity and quality, with associated impacts on agriculture, health; increases in climatic events; coastal erosion and inundation as a result of extreme events and sea level rise" (Campbell, 2012, pp. 64–65). The impact and effects of climate change continue to challenge and pressure island economies, habitats, and the livelihoods of people in the region giving rise to a broad spectrum of newly arising economic, social and cultural problems (Boege, 2018, p. 1).

The vision for a renewed story through *Re-thinking the Household of God* in the Pacific was mandated in 2013 at the Pacific Conference of Churches (PCC) general assembly. The impetus for a renewed vision for the household was the growing discomfort and unease with the destructive economic and ecological impact on the relational well-being on Pasifika peoples and their environment. The vision was the culmination of work that began two decades earlier in 2001 at the global ecumenical conference organised by the World Council of Churches, The World Alliance of Reformed Churches, and the Council of European Churches on the theme of economic globalization. The 2001 conference was hosted by the Pacific Conference of Churches in Fiji on the theme *The Island of Hope: An Alternative to Economic Globalization.* The term "Island of Hope" was described as a "fitting expression of the global, ecumenical concept of the Kingdom of God in the Pacific context" (World Council of Churches, 2002, p. 7). "The Island of Hope" was an alternative vision and practice to the project of economic globalization which entails domination through an unjust economic system (World Council of Churches, 2002, p. 5). The concept prioritises relationships, celebrates quality of life and values human beings and creation over the production of things and in which spirituality, family life, traditional economy, cultural values, mutual care and respect are its main components (World Council of Churches, 2002, p. 3). The ethics of "The Island of Hope" are based on the deep respect for the whole community of life.

The 2001 global ecumenical conference affirmed and encouraged the Pacific Churches to work toward an alternative model of development that is rooted in and takes seriously the Pacific contexts and their traditional-cultural-theological moorings. Ten years later, in 2010, the Pacific Conference of Churches developed a concept document on "Re-thinking the Household of God in the Pacific." The document outlined five key issues pertinent to the region: (i) governance and leadership; (ii) development in Oceania; (iii) peace and security; (iv) climate change and resettlement; and (v) cultural and social cohesion (Pacific Conference of Churches, 2010). The charge given at the PCC general assembly was a reiteration of the call given more than 10 years prior to "rewrite a new story of development, ecumenism, and faith that is 'Pacific oriented' and that which drives an ecological and wellbeing renewal of the Pacific" (Pacific Theological College, 2021). The rewriting of this new narrative of hope is carried out against and in resistance to a historical background of colonisation and subsequent displacement of wild thinking of which embodied practices of hope lie in the recovery and reclaiming of indigenous knowledge and spirituality displaced both by colonisation and Christian missions (Winkler & Carroll, 2021; see also Bhagwan et al., 2020).

Re-thinking the household of God in the Pacific prioritizes ecological well-being as a central concept as well as the starting point from which the understanding of theology, economics, culture, and spirituality is reframed. The "household" is understood in its broadest definition and according to its etymological Greek roots, *oikos*, of which economy, ecology and ecumene share. The management of the household or the whole inhabited earth and its inhabitants (*oikoumene*) is dependent on both good economics (management of resources) and ecology (interdependence and integration of the household) (Bird et al., 2020, p. 2). The ecological crisis, as argued by Pacific indigenous scholars, is fundamentally about the whole of life and the interdependent interconnectedness between the human community, development and the environment (Bhagwan et al., 2020, p. xv). Accompanying this understanding is the conviction that the ecological crisis cannot be resolved by scientific knowledge and fiscal means alone. The contribution of indigenous and faith based ecological frameworks (knowledge, ethics and practices) are also required to finding alternative solutions (Bhagwan et al., 2020, p. xvi).

In 2017, in conjunction with the Pacific Theological College (PTC), the PCC launched the Reweaving the Ecological Mat (REM) initiative premised on the knowledge and experience that the neoliberal model of economic development is not appropriate for the Pacific (Bird et al., 2020). REM reframes the understanding of development within an *oikonomical* framework drawing on the three interrelated words: economy, ecology and ecumenical. The *oikonomical* framing aims to "reintroduce a new/old way for measuring wellbeing and wholeness"(Bird et al., 2020, p. xi). REM is constructed around three frames – theology, economics, and cultures and spirituality. Each of these themes privilege ecological well-being and the reclaiming of Indigenous epistemologies and spiritualities in reconnecting people with the *vanua* (land). In this reframing "relationality is central to such measurement, as well as the full recognition of and respect for the balance and harmony of all diverse interwoven strands of life" (Bird et al., 2020, p. xi). Wholeness or unity is not possible without relationality. REM is wholistic in its approach. It brings together the strands of theology (Christian faith), culture and indigenous spirituality and economics into conversation, allowing each to inform, resource and enrich. In a context where indigenous epistemologies and spiritualities have been displaced by a colonial Christian narrative, this integrated approach presents a challenge to the region, in particular, to the churches who are generally reluctant to critique their received Christian traditions and theologies. Decolonising educational processes thus becomes key (Winkler & Carroll, 2021, p. 56).

4.4 How Can We Shape the Future? – Perspectives of Hope and Sustainability in Education

If religious education wants to convey its social relevance and initiate change processes in the ecological crisis from the sources of religions, it requires strenuous processes of conversion and redirection. Against this horizon, it is intended to highlight the resource that religious education can represent for sustainable development in educational systems.

4.4.1 Reweaving the Ecological Mat

Reweaving the Ecological Mat – this term from the Pacific can be used to describe the direction in which theology, culture, and spirituality as well as economy can be re-imagined and in which the necessary realization flows that the ecological crisis cannot be resolved by scientific information and fiscal means alone. In this context religious education can provide twofold by: identifying responsibility and enabling critical hope from religious traditions and their ethical bodies of knowledge.

Humans are woven into a network of life and mandated with *responsibility* (Lévi-Strauss, 1968, p. 52f). This responsibility involves a worldview and a sense of sustainability that have enabled indigenous cultures to live sustainably for millennia. Bringing their knowledge and wisdom into dialogue with religious education can become a healing re-reading of societal as well as theological power traditions. This also includes to recognizing that there are deep emotional consequences of the ecological crises. It is the existential sense of loss that people and in particular young people all over the world express as emotional pain about the destruction of nature and the exploitation of natural resources. Studies show that 75% of young people and teachers show marked pessimism about the future, and half of both learners and teachers were "'sustainability-aware but disillusioned'" (Grund & Brock, 2019, p. 9). From the combination of high awareness of the problem and still clear pessimism about the future, it follows that ESD must not stop at simply providing information. In particular, it requires *critical hope* (Ojala, 2017), that is, awareness of the real magnitude of the problem while knowing ways to address the problem and one's own and community's opportunities for participation.

4.4.2 Religious Education for a Renewed Being-in-the-World

Against this horizon, the strings are to be brought together in conclusion and opened up for religious education in the sense of a renewed being in the world. To make the elaborated theological, postcolonial, and indigenous perspectives productive for (religious) education and its response to the ecological crisis, a dual attempt has been undertaken in this paper:

On the one hand, the entanglements of Christian missionary societies in Europe's colonial structures of domination, which helped legitimize the exploitation of natural resources, have been deconstructed. This also includes perceiving that in the name of Christianity, the suppression of Indigenous forms of knowledge and knowing, which contain a treasure of care for nature and forms of sustainable living, have been legitimized (World Council of Churches, 2012). It is about the realization that the ecological crisis is a product of colonization.

On the other hand, the engagement with these historical lines of development, however, opens up the opportunity to re-engage Indigenous narratives of hope in religious education processes and to make their alternative world relations constructive, which is fed by a dialogue with Indigenous epistemology and spirituality (Lam et al., 2020). Indigenous communities have long been raising their voices to share their knowledge in the ongoing fight against climate change and the growth society. "We want to show the world what it really means to live sustainably. No one understands sustainability better than indigenous peoples." (Rio+20 UN Conference). By means of their local practices and initiatives, they share in educational projects the importance they attribute to cultural values such as respect for nature, sharing in a community, and local traditional knowledge for a healthy and resilient society and a sustainable way of life (cf. Indigene Wege, 2020). The reframing and reimagining of Indigenous narratives of hope, in dialogue with Christianity, can enable hopeful, meaningful, and ethical action rooted in solidarity, responsibility, and mindfulness for all life. Engaging Indigenous peoples in religious education processes means taking the insights of indigenous theological reflections as narratives of hope and bringing their indigenous visions of a full, good, and abundant life and their spiritual as well as theological reflections into dialogue with students (Tendis, 2020; World Council of Churches, 2012).

As part of general education, religious education has an important task in terms of transformative learning by critically discussing social and global developments and enabling an examination of questions of value and meaning from the sources of the religions (Hannam, 2019; Antal, 2018; Deane-Drummond et al., 2017). Religious education can thus open up a space in which anxieties and hopes, questions and doubts, insights and visions can be shared, but also fundamental questions can be addressed to the Christian faith: What narratives of God, of humankind and of creation can help to reshape the present in such a way that a good future for all life is possible? And what does it mean as a worldwide Christianity to take responsibility against the destroying of creation? On the basis of this educational dialogue of contexts it is obvious that more than ever there is a need for a "plurality of world references, which today has to take the place of the one-dimensionality of the traditional rationalism" (Cassirer, in: Paetzold, 1995, p. 104; Cassirer, 1923–1929).

References

Albrecht, G. (2006). Solastalgia: Environmental damage has made it possible to be homesick without leaving home. *The Free Library*. 2006 Alternatives, Inc. 19 May, 2021. https://www.thefreelibrary.com/Solastalgia%3a+environmental+damage+has+made+it+possible+to+be+homesick...-a0161545303

Albrecht, G. (2019). *Earth emotions: New words for a new world*. Cornell University Press.

Albrecht, G., Sartore, G.-M., Connor, L., et al. (2007). Solastalgia: The distress caused by environmental change. *Australasian Psychiatry, 15*(1), S95–S98. https://doi.org/10.1080/10398560701701288

Aly, G. (2021). *Das Prachtboot. Wie Deutsche die Kunstschätze der Südsee raubten*. Fischer Verlag.

Anderson, B. (2006). Becoming and being hopeful: Towards a theory of affect. *Environment and Planning: Society and Space, 24*, 733–752. https://doi.org/10.1068/d393t

Antal, J. (2018). *Climate church, climate world: How people of faith must work for change*. Rowman & Littlefield.

Armstrong McKay, D. I., Staal, A., Abrams, J. F., Winkelmann, R., Sakschewski, B., Loriani, S., Fetzer, I., Cornell, S. E., Rockstrom, J., & Lenton, T. M. (2022). Exceeding 1.5 degrees C global warming could trigger multiple climate tip-ping points. *Science, 377*(6611), eabn7950. https://doi.org/10.1126/science.abn7950

Bertelmann, B., & Heidel, K. (2018). Leben im Anthropozän: Christliche Perspektiven für eine *Kultur der Nachhaltigkeit,* oekom Verlag.

Bhagwan, J., Huffer, E., Koya-Vaka'uta, F. C., & Casimira, A. (Eds.). (2020). *From the deep: Pasifiki voices for a new story*. Pacific Theological College.

Bird, C., Saiki, A., & Ratunabuabua, M. (2020). *Reweaving the ecological mat framework: Toward and ecological framework for development*. Pacific Theological College.

BMUV, & UBA. (2022). Bundesministerium für Umwelt, Naturschutz, nukleare Sicherheit und Verbraucherschutz (BMUV) & Umweltbundesamt (UBA). (2022). Zukunft? Jugend fragen! – 2021: Umwelt, Klima, Wandel – was junge Menschen erwarten und wie sie sich engagieren. Bundesministerium für Umwelt, Naturschutz, nukleare Sicherheit und Verbraucherschutz (BMUV): https://www.bmuv.de/fileadmin/Daten_BMU/Pools/Broschueren/zukunft_jugend_fragen_2021_bf.pdf

Boege, V. (2018). Climate change and conflict in Oceania: Challenges, responses and suggestions for a policy-relevant research agenda. *Toda Peace Institute Policy Brief (No. 17)*, 1.

Brock, A., & Holst, J. (2022). Schlüssel zu Nachhaltigkeit und BNE in der Schule: Ausbildung von Lehrenden, Verankerung in der Breite des Fächerkanons und jenseits der Vorworte (Kurzbericht des Nationalen Monitorings zu Bil-dung für nachhaltige Entwicklung). https://doi.org/10.17169/refubium-36094

Campbell, J. R. (2012). Climate-induced community relocation in the Pacific: The meaning and importance of land. In J. McAdam (Ed.), *Climate change and displacement: Multidisciplinary perspectives* (pp. 57–80). Oxford Hart Publishing.

Cassirer, E. (1923–1929). *Philosophie der symbolischen Formen*, 3 Bde. Berlin.

Chimee, I. N. (2018). *African historiography and the challenges of European periodization: A historical comment*. https://trafo.hypotheses.org/11518

Deane-Drummond, C., Bergmann, S., & Vogt, M. (Eds.). (2017). *Religion in the Anthropocene*. Cascade Books.

Döbler, K. (2021). *Dein ist das Reich*. Claasen Verlag.

Doran, C. (2017). *Hope in the age of climate change: Creation care this side of creation*. Cascade Books.

Ellis, N., & Cunsolo, A. (2018). Hope and mourning in the Anthropocene: Understanding ecological grief. *The Conversation*. Retrieved from https://theconversation.com/hope-and-mourning-in-the-anthropocene-understanding-ecological-grief-88630

Franzen, J. (2019). *What if we stopped pretending?* Retrieved from https://www.newyorker.com/culture/cultural-comment/what-if-we-stopped-pretending

Goodman, M. K., McNatt, M. B., & Boykoff, M. (2022). Communicating climate change in the Anthropocene: The dynamic cultural politics of climate change news coverage and social media around the world. In A. Hansen & R. Cox (Eds.), *International handbook of environment and communication* (2nd ed.). Routledge.

Gräsel, C. (2010). Stichwort: Transfer und Transferforschung im Bildungsbereich. *Zeitschrift für Erziehungswissenschaft, 13*(1), 7–20. https://doi.org/10.1007/s11618-010-0109-8

Green Colonialism. https://earth.org/green-colonialism/; https://worldcrunch.com/tech-science/green-colonialism-the-new-face-of-environmental-hypocrisy; https://ideas4development.org/en/green-colonialism-western-outlook/; https://www.euractiv.com/section/arctic-agenda/news/nordic-countries-set-up-sami-reconciliation-commissions-to-investigate-indigenous-injustices/

Gruber, J. (2018). Wider die Erinnerung. Zur postkolonialen Kritik hegemonialer Wissenspolitiken in der Theologie, in: Andreas Nehring – Simon Tielesch (Hg.) (2013): *Postkoloniale Theologie. Bibelhermeneutische und kulturwissenschaftliche Beiträge (Religionskulturen, Band 11)*, Stuttgart, 23–37.

Grund, J., & Brock, A. (2018). Bildung für nachhaltige Entwicklung in Lehr-Lernsettings: – Quantitative Studie des nationalen Monitorings – Befragung junger Menschen. https://www.ewi-psy.fu-berlin.de/einrichtungen/weitere/institut-futur/Projekte/Dateien/Grund_-J__-Brock_-A__2018__BNE_in_Lehr-Lernsettings_Quantitative-Studie____Befra-gung_junger-Menschen.pdf

Grund, J., & Brock, A. (2019). Why we should empty Pandora's box to create a sustainable future: Hope, sustainability, and its implications for education. *Sustainability, 11*(3). https://doi.org/10.3390/su11030893

Grund, J., & Brock, A. (2022). Formale Bildung in Zeiten von Krisen – die Rolle von Nachhaltigkeit in Schule, Ausbildung und Hochschule. Kurzbericht des Nationalen Monitorings zu Bildung für nachhaltige Entwicklung (BNE) auf Basis einer Befragung von > 3.000 jungen Menschen und Lehrkräften. Institut Futur, Freie Universität Berlin. https://doi.org/10.17169/refubium-36890

Guyatt, R. (2020). Kierkegaard in the Anthropocene: Hope, philosophy, and the climate crisis. *Religions, 11*(279), 1–12. https://doi.org/10.3390/rel11060279

Habermas, R., & Hölzl, R. (Hg.). (2014). Mission global. Eine Verflechtungsgeschichte seit dem 19. *Jahrhundert,* Böhlau Verlag Köln.

Hamilton, C. (2017). *Defiant Earth. The fate of humans in the Anthropocene*. Polity Press.

Hannam, P. (2019). *Religious education and the public sphere*. Routledge.

Head, L. (2016). *Hope and grief in the Anthropocene* (Kindle ed.).

Holst, J. (2022). Nachhaltigkeit & BNE in der Beruflichen Bildung: Dynamik in Ordnungsmitteln, Potentiale bei Berufen, Lernorten und in der Qualifizierung von Ausbildenden (Kurzbericht des Nationalen Monitorings zu Bildung für nachhaltige Entwicklung). https://doi.org/10.17169/refubium-35827

Holst, J., & Singer-Brodowski, M. (2022). Nachhaltigkeit & BNE im Hochschulsystem: Stärkung in Gesetzen und Zielvereinbarungen, ungenutzte Potentiale bei Curricula und der Selbstverwaltung (Kurzbericht des Nationalen Monitorings zu Bildung für nachhaltige Entwicklung). https://doi.org/10.17169/refubium-35828

Indigene Wege. (2020). *Wandel-Erfahrungs-Geschichten indigener Gemeinschaften zu den nachhaltigen Entwicklungszielen (SDG) mit Begleitheft mit Handreichungen zu den 17 Nachhaltigkeitszielen für die Sekundarstufe*. Institut für Ökologie und Aktions-Ethnologie. https://www.infoe.de/projekte/indigene-und-sdgs

Kirchhoff, J. (1999). *Räume, Dimensionen, Weltmodelle. Impulse für eine andere Naturwissenschaft*. Diederichs.

Lam, D., Hinz, E., Lang, D., Tengö, M., von Wehrden, H., & Martín-López, B. (2020). Indigenous and local knowledge in sustainability transformations research: A literature review. *Ecology and Society, 25*(1), 3. https://doi.org/10.5751/ES-11305-250103

Lephakga, T. (2015). The history of conquering of the being of Africans through land dispossession, epistemicide and proselytization. *Studia Historiae Ecclesiasticae, 41*(2), 145–163.

Lévi-Strauss, C. (1968). *Das wilde Denken*. Suhrkamp.

Marlon, J. R., Bloodhard, B., Ballew, M. T., Rolfe-Redding, J., Roser-Renouf, C., Leiserowitz, A., & Maibach, E. (2019). How hope and doubt affect climate change mobilization. *Frontiers in Communication, 4*, 1–15. Retrieved from https://www.frontiersin.org/articles/10.3389/fcomm.2019.00020/full

Meyer, M. (2013). Holographic epistemology: Native common sense. *China Media Research, 9*(2), 94–101.

Ojala, M. (2012). Hope and climate change: The importance of hope for environmental engagement among young people. *Environmental Education Research, 15*(5), 625–642. https://doi.org/10.1080/13504622.2011.637157

Ojala, M. (2017). Hope and anticipation in education for a sustainable future. *Futures, 94*, 76–84. https://doi.org/10.1016/j.futures.2016.10.004

Orr, D. W. (1994). *Earth in mind: On education, environment, and the human Prospect*. Island Press.

Ortiz, D. A. (2020). *Is it wrong to be hopeful about climate change?* Retrieved from https://www.bbc.com/future/article/20200109-is-it-wrong-to-be-hopeful-about-climate-change

Pacific Conference of Churches. (2010). *Re-thinking the household of God: Towards a region of sufficiency, solidarity, inclusiveness and participation*. Pacific Conference of Churches.

Pacific Theological College. (2021). *University transition of the Pacific Theological College: Forging a whole of life pathway in Pasifika education*. P. T. College.

Paetzold, H. (1995). Ernst Cassirer – Von Marburg nach New York. *Eine philosophische Biographie*, Wiss. Buchgesellschaft Darmstadt, 104.

Pittl, S. (Hg.). (2018). Theologie und Postkolonialismus: Ansätze – Herausforderungen – Perspektiven, Ansätze – Herausforderungen – Perspektiven, Regensburg Verlag Friedrich Pustet.

Ratschiller, L., & Wetjen, K. (Hg.). (2018). Verflochtene Mission Perspektiven auf eine neue Missionsgeschichte, Böhlau Verlag Köln.

Raygorodetsky, G. (2017). *The Archipelago of Hope: Wisdom and resilience from the edge of climate change* (Kindle ed.).

Raygorodetsky, G. (2018). Indigenous peoples defend Earth's biodiversity – But they're in danger. *National Geographic* Retrieved from https://www.nationalgeographic.com/environment/article/can-indigenous-land-stewardship-protect-biodiversity-

Robinson, T. (2020). Reimagining Christian Hope(lessness) in the Anthropocene. *Religions, 11*(192), 1–11.

Steffen, W., Richardson, K., Rockstrom, J., Cornell, S. E., Fetzer, I., Bennett, E. M., Biggs, R., Carpenter, S. R., de Vries, W., de Wit, C. A., Folke, C., Gerten, D., Heinke, J., Mace, G. M., Persson, L. M., Ramanathan, V., Reyers, B., & Sor-lin, S. (2015). Planetary boundaries: Guiding human development on a changing planet. *Science, 347*(6223), 1259855. https://doi.org/10.1126/science.1259855

Sugirtharajah, R. S. (2013). Eine postkoloniale Untersuchung von Kollusion und Konstruktion in biblischer Interpretation, in: Andreas Nehring – Simon Tielesch (Hg.) *Postkoloniale Theologie. Bibelhermeneutische und kulturwissenschaftliche Beiträge* (Religionskulturen, Band 11), Stuttgart, 123–144.

Tendis, N. (Ed.). (2020). *Roadmap for congregations, communities and churches for an economy of life and ecological justice*. World Council of Churches. https://www.oikoumene.org/resources/documents/roadmap-for-congregations-communities-and-churches-for-an-economy-of-life-and-ecological-justice

Torre, M. A. D. L. (2017). *Embracing hopelessness*. Fortress Press.

Treanor, B. (2018). Hope in the age of the Anthropocene. *Analecta Hermeneutica, 10*, 1–22.

UNESCO, & Deutsche UNESCO Kommission (DUK). (2021). *Bildung für nachhaltige Entwicklung: Eine Roadmap.* https://unesdoc.unesco.org/ark:/48223/pf0000379488

UNHCR. (2020). *Land-grabbing in Asia displaces indigenous peoples and destroys environment, says UN rights expert*. Retrieved from https://www.ohchr.org/EN/NewsEvents/Pages/DisplayNews.aspx?NewsID=26213&LangID=E

Vaai, U. L. (2019). "We are therefore we live" Pacific eco-relational spirituality and changing the climate change story. *Toda Peace Institute Policy Brief (No. 56),* 1–14.

Vaai, U. L., & Casimira, A. (Eds.). (2017). *Relational hermeneutics: Decolonising the mindset and the Pacific Itulagi*. The University of the South Pacific & The Pacific Theological College.

Winkler, K., & Carroll, S. (2021). Solastalgia and Hope: Theological and postcolonial perspectives from Europe and the Pacific responding to the ecological crisis in (religious) education, Religionspädagogische Beiträge. *Journal for Religion in Education, 44*(2), 49–60. https://doi.org/10.20377/rpb-150

World Council of Churches. (2002). *The Island of Hope: A Pacific alternative to economic globalization. Dossier 7*. WCC Publications.

World Council of Churches. (2012). *Statement on the doctrine of discovery and its enduring impact on Indigenous Peoples*. https://www.oikoumene.org/resources/documents/statement-on-the-doctrine-of-discovery-and-its-enduring-impact-on-indigenous-peoples

Kathrin Winkler is a Professor of Religious Education and Religious Studies at the Lutheran University of Nuremberg, Germany. She is the director of the Research Institute for Theology & Society and focuses on the topics of sustainability, climate justice and postcolonialism. In her research topics she also deals with the significance of culture, religion, and migration in the context of educational processes in immigration societies.

Chapter 5
Beyond the Dhamma: Cultivating More-Than-Human Sustain-abilities Through Buddhist Stories in Thailand

Ignasi Ribó

5.1 Jātakas in Traditional Buddhist Education

The *jātakas* are basically stories that recount the previous lives of the Buddha, before he was born during the sixth or fifth century BCE as Siddhārtha Gotama, attained enlightenment (*bodhi*), and was freed from worldly suffering (P. *nibbāna*), becoming thereafter the inspiration and spiritual leader of a growing community of followers (*sangha*) and eventually of the religion known as Buddhism. In this broad sense, *jātakas* are found in a wide variety of texts, from ancient Buddhist scriptures to modern compositions and compilations (Appleton, 2010). Most likely, they have also been continuously told and retold in many different ways as part of the oral traditions of Buddhist cultures, both in the Mahāyāna and Theravāda schools, as well as in other lesser known teachings and practices (Cowell, 1895).

In the Theravāda tradition, which has been the dominant in Southeast Asia, including Thailand, the classical compilation of *jātakas* is the *Jātakatthavaṇṇanā*, which was most likely composed by Sri Lankan monks in the Pāli language around the fifth or sixth century CE (von Hinüber, 1998). This collection of 550 *jātakas* has a carefully planned structure, where every story of a previous birth (P. *atīta-vatthu* or "story of the past") is framed by a preface, a conclusion, and one or more verses (*gāthā*), all of which are told by the Buddha himself in his last incarnation (Cowell, 1895). From the analysis of this intricate and layered structure, modern scholarship suggests that the whole collection was arranged in order to provide a narrative illustration to already existing canonical verses (Oldenberg, 1912). The Buddhist monks

I. Ribó (✉)
Linguistics, Literature and Language Education for Sustainability (LLLES), Mae Fah Luang University, Chiang Rai, Thailand
e-mail: ignasi.rib@mfu.ac.th

S. Windsor, O. Franck (eds.), *Intersections of Religion, Education, and a Sustainable World*, https://doi.org/10.1007/978-3-031-81809-7_5

who most likely authored the *Jātakatthavaṇṇanā* would have gathered a wide variety of older folktales from Sanksrit and other sources and retold them as stories of Buddha's previous births (Rhys-Davis, 1917). In this sense, it seems clear that most *jātaka* tales do not have a Buddhist origin, even if "it is impossible to treat *jātakas* as non-Buddhist or even pre-Buddhist for the simple reason that these stories are now–and have been for over 2,000 years–a popular part of Buddhism" (Appleton, 2010, p. 10).

The central role of these stories in Buddhist education, particularly in the Theravāda tradition, is undeniable:

> These legends were also continually introduced into the religious discourses which were delivered by the various teachers in the course of their wanderings, whether to magnify the glory of the Buddha or to illustrate Buddhist doctrines and precepts by appropriate examples, somewhat in the same way as mediaeval preachers in Europe used to enliven their sermons by introducing fables and popular tales to rouse the flagging attention of their hearers (Cowell, 1895, p. vi).

In contrast to the canonical texts, where Buddhist doctrine and philosophy is often expressed in an abstruse and technical language, the *jātakas* present the central ideas and worldview of the *sangha* through the means of popular narratives that can appeal to common people, in monastic settings as well as amongst the laity, inspiring them to follow the path of the Buddha (Pierce, 1969). The pedagogical function of these stories is evident from the widespread depiction of scenes and motifs drawn from *jātakas* in the paintings and sculptures that decorate Buddhist temples throughout Asia (Gombrich, 2006). Even in modern times, these stories continue to play an important role in the moral education of children in countries like Thailand, where the adoption of Western educational models and technologies has not fully undermined the pedagogical and ideological principles derived from the old tradition of temple education (Singhsa, 2011).

As religious parables, the *jātakas* have generally been used in the Theravāda tradition to provide vivid examples of the moral principles that are supposed to guide the life and actions of good Buddhists, whatever their status or condition might be. In this context, whether in a temple or in a modern classroom, the stories are presented as a way to support the *Dhamma* (S. *Dharma*), the learning of Buddhist morality, as it has been interpreted by monks (P. *bhikkhu*/*bhikkhunī*) and other religious authorities. Regardless of their origin, therefore, the stories of the previous births of the Buddha are generally interpreted and taught to pupils as an illustration of the moral virtues, the ten *pāramitās*, that allowed the *bodhisatta* to progress through multiple nonhuman and human reincarnations (*samsāra*) in his path to enlightenment as the Buddha. Of course, the fundamental lesson that monks and teachers in this tradition aim to instill in their pupils is the need for everyone to follow this same path in order to attain buddhahood, which is presented as "the ultimate goal of perfection for all sentient beings" (Chen, 2003, p. 15).

This dogmatic pedagogy is certainly problematic, and perhaps even counterproductive, when considered in the context of education for sustainability, particularly when critical or emancipatory aims are an integral part of the meaning of sustainability (Kahn & Humes, 2009; Wals, 2012; Wals et al., 2008). Such problems are generally disregarded by non-Buddhists who find in Buddhism an exotic alternative to the cultural patterns, often derived from the Christian tradition, that are dominant in Western education. By decontextualising the *Dhamma* and cherrypicking on generic virtues like compassion or mindfulness (Keiser & Adarkar, 2015), it is certainly possible to articulate pedagogies inspired in Buddhism that have a certain appeal in the highly secularised and materialistic Western world. However, orientalist undertakings (King, 2013; Said, 1979) of this kind tend to ignore the extent to which Buddhist education has been historically, and even today still is, tied up with authoritarian and conservative forms of education in those places where it is the dominant religion.

The dogmatic and ideological aspects of the *jātakas* are already obvious in the way these ancient folktales were framed in the earliest texts of the Pāli canon, where the monastic authors imposed on them "a master narrative which seeks to provide both anchorage and closure to units which may otherwise be open-ended and capable of being 'consumed' differently" (Chakravarti, 1993, p. 50). In Southeast Asia, and particularly in Thailand, the *jātakas* have also been instrumental in anchoring a certain model of political and moral authority, founded on a hierarchical and centralised conception of the social order (Jory, 2016; Ribó, 2017). The modernisation of the country, particularly since the reign of Chulalongkorn (Rama V), resulted in the replacement of traditional monastic education with a centralised and bureaucratic system loosely modelled on the West (Singhsa, 2011). This push towards modernity also tended to weaken the role of the *jātakas* as "the principal conduit of a conception of authority and social hierarchy" (Jory, 2002). And yet, even as Thai society has continued to modernise and has become increasingly secularised, these stories are still used by monks in temples, by teachers in schools, and by parents at home, to educate children in the eightfold path of the Buddhist *Dhamma* (right view, right resolve, right speech, right conduct, right livelihood, right effort, right mindfulness, and right concentration).

5.2 Environmental Ethics in the Jātakas

The widely held association of Buddhism with environmental ethics is, to a large extent, a product of Western environmentalism (Callicott, 2008). In his damming historical appraisal of the Judaeo-Christian roots of the environmental crisis, Lynn White Jr. remarked that Zen Buddhism "conceives of the man-nature relationship as very nearly the mirror image of the Christian view" (1967, p. 1206), justifying the growing interest in Buddhist environmental ethics and practices by Western activists, artists, and academics from the 1960s onwards. This is not to say, of course, that Buddhism, like many other religions, does not include traditions and doctrines that effectively support pro-environmental beliefs, values, and actions (Thathong, 2012). However, "there are reasons to doubt the claim that that the Buddhist teachings of non-self, conditioning arising, and emptiness amount to anything like an ecological view of nature" or even that "Buddhism is environmentally friendly" (James, 2013, p. 602).

First of all, we should not forget that Buddhism, far from being a unitary belief system, constitutes a complex and multifaceted phenomenon, just as ecology or environmental ethics. It is doubtful, for instance, that early Buddhists afforded value to the natural environment beyond its role in supporting the meditative practices that allowed them to escape from worldly suffering and attain enlightenment (Schmithausen, 1997). Similarly, while early doctrine regarding non-violence (P. *avihiṃsā*), retribution (P. *kamma*), kindness (P. *mettā*), and compassion (P. *karuṇā*) might have supported a more humane treatment of sentient nonhuman animals in Buddhist communities, particularly when compared with other world religions (Sahni, 2008), it did so to a limited degree, and in any case much less than amongst contemporary Jains (Finnigan, 2017).

What is important to understand is that the worldview of early Buddhism, which was later developed through monastic communities in South Asia and shaped the foundations of the Theravāda orthodoxy, is profoundly anthropocentric. While there is a recognition that both human and nonhuman beings are tied up in the same cycle of death and rebirth (*samsāra*), from which a certain sense of responsibility and relationality is derived, the fundamental teachings of Buddhism place all nonhumans on a lower scale of being, insofar as they are "more unfavourably disposed to the possibility of liberation for three basic reasons; they are inferior to humans existentially, morally and intellectually" (Harris, 2000, p. 119). Thus, for most Buddhists, being reborn as a nonhuman animal is a particularly unfortunate event, which is often interpreted as a punishment for past misdeeds and, in any case, a step back in the moral progress towards buddhahood. The result is a Buddhist version of the Great Chain of Being, not so different from its Christian counterpart (Lovejoy, 1936), which is often represented on Theravāda temples, with the various realms of ghosts, demons, animals, humans, divinities and, always at the top, the Buddha, hierarchically layered to display the right path to salvation.

This anthropocentric ontology is much less pronounced in the *jātakas*, where the *bodhisatta* is often presented in his previous life as a nonhuman animal, or even a tree, in order to provide examples of virtuous actions that can be interpreted, in modern terms, as environmentally friendly (Sahni, 2012). There are also many other nonhuman animals characterised in these stories, both in a central role or as part of the background (Singh, 2015). In general, though, these tend to be heavily anthropomorphised and are used, like in Aesop's and other fables, to represent human virtues and vices, showing little attention to the actual experience or perspective of the real living animals (Waldau, 2000). Moreover, the *jātakas* also include a considerable number of nonhuman beings intimately connected to the natural world, such as nats, yakkhas, demons, and other spirits, which seem to emerge from pre-existing oral traditions (Harris, 2000).

Given the heterogeneous and layered origin of the *jātakas*, it is very difficult to discern the source of the various elements that today might be recognisable in these stories as promoting sustainability or environmental ethics. In particular, it is not clear whether these values reflect the ontological perspectivism (Viveiros de Castro, 1998) of pre-Buddhist animistic beliefs, the ascetism of the "hermit strand" of early Buddhism (Schmithausen, 1997), or the anthropocentric and humanistic virtue ethics that characterises Buddhism as a global religion. Because the stories have been overlaid, generation after generation, with the worldview of the Buddhist *sangha*, with its emphasis on liberation and right moral conduct, using them in the context of education for sustainability represents a real challenge. Unless our explicit aim is to educate on the basis of religious beliefs, values, and practices, as traditional or modern Buddhist education (P. *sikkha*) would attempt to do (Chansomsak & Vale, 2008), we have to recognise that the ontological and axiological foundations of the *Dhamma* can only encourage, at best, weakly anthropocentric forms of engagement with the environment, which may be criticised from a posthumanist perspective as "stewardship framings" (Taylor, 2017). To develop non-anthropocentric pedagogies that result in more sustainable forms of relationality between humans and nonhumans, it might be necessary to go beyond the *Dhamma* and find new ways of incorporating environmentally relevant stories like the Buddhist *jātakas* in the classroom. In what follows, I discuss some aspects of a recent experience in the context of Thailand that attempts to move education for sustainability in this direction.

5.3 Beyond Humanism

During the second semester of the academic year 2022–23, I taught a new elective course in *Environment, Literature and Culture*, as part of the Bachelor of English at the School of Liberal Arts of Mae Fah Luang University, in the north of Thailand. Instead of conceiving this course in terms of the liberal-humanistic model of education, which underlies the hermeneutic methods generally used in literary pedagogy, I tried to develop an "unprecedented" (Garrard, 2017) ecocritical curriculum based on emerging posthumanist ontologies and pedagogies (Snaza et al., 2014; Snaza & Weaver, 2015). Hence, my aim was not to "cultivate humanity" (Nussbaum, 1997) or "cultivate sustainability" (Bartosch, 2021), but rather to "face humanity" (Todd, 2015) and undermine the "anthropological machine" of humanism (Agamben, 2004), which is constantly attempting to separate the human from the nonhuman and constitutes, in my opinion, the fundamental obstacle for the emergence of more sustainable forms of cohabitation between fellow beings-in-the-world (Ribó, 2012, 2022).

With this course, therefore, I tried to encourage students to develop an embodied, affective, experienced, relational sense of becoming-together with other co-inhabitants of the Earth, by putting in place situated and dialogical pedagogical invitations that aimed at cultivating more-than-human *sustain-abilities*. These are the crucial abilities that students will need to sustain themselves and to sustain others, including both humans and nonhumans, in the face of the challenges and risks of the unfolding Anthropocene (Ribó, 2023b). In particular, they include *vulner-ability* (being able to recognise one's own fragility and exposure to harm), *attend-ability* (being able to attend to another's harm, to pay attention or to listen to the experience of the other in its own terms), and *response-ability* (being able to respond to another's harm by acting and taking care of the other).

Beyond my role as designer and instructor, I approached each meeting of the course as an emergent "literacy situation" (Snaza, 2019), where various human agencies (12 students and myself), but also a myriad of nonhuman actants (both animate and inanimate), were assembled and produced unique and ephemeral literacy events through intra-active entanglements (Barad, 2007). By taking part in these pedagogical invitations, both the students and myself were open to experience moments of "bewilderment" (Snaza, 2019), becoming perceptually, imaginatively, and emotionally affected by the literacy event in such a way that anthropocentric subjective formations, ultimately stemming from the "anthropological machine" of humanistic education (Agamben, 2004; Lewis & Kahn, 2010; Snaza, 2013), could be disrupted and eventually overcome (Ribó, 2023a).

The pedagogical invitations of the course included a considerable number of outdoor activities and encounters with nonhumans, as well as the reading and discussion of a broad selection of literature, from prose fiction and poetry to scientific, philosophical, and journalistic texts engaging in various ways with the environment. Amongst these texts, I asked students to read two little known *jātakas*, "The tigress" (Āryaśūra, 1983) and "Others sow" (Cowell, 1901), drawn from Mahāyāna and Theravāda sources. In the following sections, I will briefly summarise both of these stories, contextualise them, and explain how I tried to use them in class to encourage the development of sustain-abilities on the part of those students who participated in the course.

I should mention that these students had all been brought up in contexts strongly shaped by Theravāda Buddhism and were quite familiar with the narrative, symbolic, and ideological content of the *jātakas*, even if none of them had ever heard or read these two stories in particular. In an increasingly secularised society like Thailand, young people often dismiss Buddhist or other religious tales, which are generally used by teachers, monks, and parents to convey a moralising and patronising discourse. In my class, therefore, I explicitly avoided treating the *jātakas* as examples of virtue or instruments of moral or religious education. Instead, I reframed them in the context of a non-anthropocentric and dialogic pedagogy (Murphy, 1991; Tassoni, 1998), by downplaying their religious or dogmatic dimensions and inviting students to engage with the stories in their own terms, as open and contentious propositions. My aim was to encourage the assemblage of a literacy situation that promoted moments of "bewilderment" (Snaza, 2019) and allowed students to reconsider and perhaps put into question, but not necessarily to reject, a worldview nourished by years of anthropocentric education.

5.4 The Tigress

> In one of his previous births, the bodhisatta was a revered teacher, living a life of virtue and practicing meditation in the forest. One day, walking up a mountain trail with one of his disciples, he came upon a tigress trapped below a ravine with her newborn cubs. Roaring of hunger, the tigress seemed about to eat her own offspring. The bodhisatta was moved to great compassion and asked his disciple to go fetch some meat for the tigress. As time passed and his disciple was not coming back, the bodhisatta threw himself over the cliff, giving up his own life so that the tigress could eat. When finally the disciple returned empty-handed and saw his master being devoured by the starving tigress, he felt sorrow, but also admiration for such an act of selflessness and compassion for all sentient beings (summarised from Āryaśūra, 1983).

This story, probably a folktale of pre-Buddhist origin, has been recounted in various ways, particularly in the Mahāyāna tradition, with the *bodhisatta* being presented as the man who sacrifices his life so that the hungry tigress and her cubs may live. The text I used for the course was originally written in Sanskrit by Āryaśūra in the fourth century CE, as part of a collection called *Jātakamālā*, translated into English by Joseph Speyer in 1895 and modernised in a recent edition based on a Tibetan translation from the eighth century (Āryaśūra, 1983). The story, which has also inspired various shrines in Nepal and China, elaborates a recurrent theme in many of the *jātakas*, the virtue of self-sacrifice as a form of detachment from the world and compassion towards all living creatures (Pierce, 1969).

In general, the idea of virtue is a profoundly anthropocentric one, founded on the idealisation of human dignity and the associated notions of moral duty, both in relation to oneself and the community. Stories of self-sacrifice, where the *bodhisatta* gives away his wealth, wife, children, limbs, or even life (P. *pañca-mahāpariccāga*) are generally presented by Buddhist teachers as examples of virtuous moral actions (Appleton, 2010). But even in the context of a metaphysic of reincarnation (*samsāra*), stories of this kind are highly problematic, given that the sacrifice of one's own life can only be justified within a humanistic ethical framework as an expression of the ultimate virtue of an idealised figure like the Buddha (Sheravanichkul, 2008).

In class, I explicitly asked students to read the story laying aside all these religious justifications and assuming that the protagonist was a normal human being, perhaps a foreigner named John Smith hiking in the forests of Laos. At the same time, I avoided giving them any alternative framework to interpret or make sense of this man's self-sacrifice. The story produced a strong impact on students and, in spite of my attempts at decontextualising it, some of them quickly sought an explanation in the fact that the protagonist was the Buddha. However, I was able to reconduct the discussion by asking them directly whether they would act in the same way as the protagonist if they found themselves in this situation. Overwhelmingly, students responded in the negative, dismissing my question as absurd. But then I changed the situation by turning the trapped tiger into a human female with her babies. Predictably, when I asked them whether they would be willing to die to save this family, the responses from students were much stronger. Some of them still refused any idea of self-sacrifice, but others changed their position and admitted that they would consider it in this other situation. With the ethical dilemma clearly laid out and students actively engaged in the discussion, I could then introduce crucial questions regarding speciesism and the different moral standards that we use in relation to human and nonhuman animals. The bewilderment produced by the story and the literacy situation that it had prompted also allowed me to present important notions in environmental ethics, such as the sources of empathy and care towards human and nonhuman others (Haraway, 2008; Puig de la Bellacasa, 2017), the expanding moral circle (Singer, 1981), and the various attitudes that humans have towards the environment, such as weak and strong anthropocentrism, biocentrism, and ecocentrism (Desjardins, 2013), in a way that students could understand and engage with at a personal level.

5.5 Others Sow

> A thousand carpenters and their families built ships and emigrated from Benares to escape their debts. They reached a distant island that was only inhabited by a castaway, who lived a carefree and happy life thanks to the abundance of fruits that grew all over the land without any effort. The castaway welcomed the carpenters but warned them that the island was inhabited by spirits (P. *Amanussas*) who would tolerate the presence of humans as long as they took care of burying their excrement under the sand and did not soil the land. The carpenters settled in the island and prospered. After some time, they brewed liquor from sugar cane to celebrate their luck in finding such a bountiful place to dwell. Being drunk, they forgot the castaway's warning and relieved themselves carelessly throughout the island. On seeing this, the spirits were so enraged that they decided to bring a large wave to clean the island and destroy all its human inhabitants. One of the spirits, however, took pity on the carpenters and warned them of the danger. Half of the carpenters and their families, led by a wise man, heeded the warning and embarked on their ships to avoid disaster. But the rest of the carpenters, moved by greed and not wanting to lose their comfortable life, dismissed the warning and remained on the island. When the day set by the spirits arrived, a giant wave rose from the ocean and rolled over the land, killing all the men, women, and children who had stayed behind (summarised from Cowell, 1901).

This story (P. *samudda-vāṇija jātaka*) was also most likely an Indian folktale before being included in the Pāli compilation of the *Jātakatthavaṇṇanā*, which was first translated into English under the supervision of E. B. Cowell between 1895 and 1907 (Cowell, 1901). In this case, the Buddhist elements are less obvious and we only learn at the end that the *bodhisatta* was supposed to be the wise carpenter who takes heed of the warning from a benevolent spirit and flees the island with half of the families before the giant wave destroys everyone else. In the original text, the Pāli term used to refer to these island spirits is *Amanussa*, which generically designates nonhuman beings that are intimately connected with the natural world. In English and other modern translations, however, they are often rendered as gods, divinities, or other anthropomorphic and transcendent entities, something that contributes to cloud the fairly obvious animistic substrate of the story.

My initial intention was to use this story as part of an outdoor class dedicated to observe and experience the impact of climate change on human and nonhuman sustainability. In modern environmentalist terms, the *jātaka* can be read as an extraordinarily compelling parable of the negative impacts of human civilisation's expansion on the natural environment and the risks created by anthropogenic waste and pollution (Beck, 1992). It also provides a vivid illustration of the "precautionary principle" in the face of systemic environmental crises (Harremoes et al., 2013), by contraposing the attitudes of two of the carpenters, one of whom effectively takes action to prevent disaster while the other refuses to pay attention to the warning signs in order to continue to profit from the available natural resources.

On this occasion, however, my plans were disrupted by a severe episode of transboundary haze, which affected the north of Thailand at the end of March 2023, forcing the university to cancel all in-class and outdoor activities during several weeks, in order to protect students and employees from the pernicious health effects of particulate matter (PM2.5) in the atmosphere. Under these circumstances, the discussion of the *jātaka* in the online classroom focused less on the "slow violence" of climate change (Nixon, 2011) and more on the immediate and tangible consequences of a crisis that directly affected students and myself, but also a myriad nonhuman others who did not have the possibility of retreating into air-filtered rooms to protect themselves from the toxic smoke emitted by thousands of man-made fires across the region. In this way, the fires and the haze itself became entangled in the literacy situation, contributing to make the anthropogenic pollution described in the *jātaka* even more relevant for these students. This seemed to encourage them to develop a sense of their own vulner-ability, but also their ability to attend and respond to the harms endured by both humans and nonhumans in the Anthropocene.

5.6 Educating Through jātakas in the Anthropocene

Through the engagement with these *jātakas* and various other activities, the literacy situation that was assembled (Latour, 2005) in class allowed students to re-examine their own religious and cultural background in ways that one student, coming back from the temple of Phra Awalokitesuan Kuan Im on the third week of the course, described as "eye-opening." More importantly, the literacy situation that emerged from these pedagogical invitations encouraged students to put into question the anthropocentric foundations of their previous education, transforming their view of the world and of themselves, and opening them up to forms of relationality that effectively took into consideration the perspective of nonhuman others. In this context, the *jātaka* texts contributed to students' experience of bewilderment and entanglement, allowing them to develop a sense of the shared vulnerabilities that bring human and nonhuman co-inhabitants together, as well as their ability to attend to these nonhuman others and to respond to their needs through acts of care.

The more-than-human entanglements shown in this kind of stories are not only essential for a sustainable future, but they can also be the foundation of pedagogies that encourage ethical engagements with others. For this purpose, however, these pedagogies should not be grounded on a dogmatic and rigid idea of virtue, which often leads to forms of education that attempt to indoctrinate students into following a pre-determined path to rightness. Rather than offering principles, norms, or examples to follow, stories provide opportunities to come face-to-face (Levinas, 2011) with human and nonhuman others in ways that, for the most part, we cannot experience in everyday life. Narrative fictions, particularly when they attempt to go beyond the human (Weik von Mossner, 2017), are perhaps one of the most powerful ways through which both children and adults engage into these relationships. It is not so much that fictions allow us to recognise and comprehend the life choices of other beings, as Nussbaum claims (1997), but that they invite us to experience imaginary and affective encounters from which the attitudes of compassion and care can emerge (Ribó, 2013).

In an epoch when human activities are causing devastating global changes that undermine the capacity of multispecies communities to sustain themselves, stories like "The tigress" or "Others sow" can contribute to forms of teaching and learning that overcome the ideology of humanism and cultivate a posthumanist or, at least, a post-anthropocentric relationship between humans and nonhumans. When presented without the metaphysical and doctrinary wrappings accumulated after centuries of religious teaching, these stories may help to erode humanistic dogmas by inviting students to critically reconsider the assumption that human life is more valuable than nonhuman life. In the contemporary classroom, ancient stories like the *jātakas* offer opportunities for dialogical and situated pedagogical invitations that go beyond the humanistic model of education (Ribó, 2023b). Such a dialogue might not teach more sustainable behaviours in a predefined, normative, or instrumental sense (Biesta, 2015; Wals, 2012; Wals et al., 2008), but it can stimulate the kind of transformative, emancipatory, and self-critical learning processes that, for many, are the fundamental aims of education for sustainability.

References

Agamben, G. (2004). *The open: Man and animal*. Stanford University Press.

Appleton, N. (2010). *Jātaka stories in Theravāda Buddhism: Narrating the Bodhisatta path*. Ashgate.

Āryaśūra. (1983). The tigress. In *The marvelous companion: Life stories of the Buddha* (pp. 2–9). Dharma Publishers.

Barad, K. (2007). *Meeting the Universe Halfway: Quantum physics and the entanglement of matter and meaning*. Duke University Press.

Bartosch, R. (Ed.). (2021). *Cultivating sustainability in language and literature pedagogy: Steps to an educational ecology*. Routledge.

Beck, U. (1992). *Risk society: Towards a new modernity*. SAGE.

Biesta, G. (2015). What is education for? On good education, teacher judgement, and educational professionalism. *European Journal of Education, 50*(1), 75–87.

Callicott, J. B. (2008). The new new (Buddhist?) ecology. *Journal for the Study of Religion, Nature & Culture, 2*(2).
Chakravarti, U. (1993). Women, men and beasts: The Jataka as popular tradition. *Studies in History, 9*(1), 43–70.
Chansomsak, S., & Vale, B. (2008). The Buddhist approach to education: An alternative approach for sustainable education. *Asia Pacific Journal of Education, 28*(1), 35–50. https://doi.org/10.1080/02188790701850063
Chen, N. (2003). Educational philosophy in humanistic Buddhism. *Hsi Lai Journal of Humanistic Buddhism= 西來人間佛教學報, 4*, 13–22.
Cowell, E. B. (1895). Preface. In E. B. Cowell (Ed.), & R. Chalmers (Trans.), *The Jātaka, or Stories of the Buddha's Former Births* (Vol. 1, pp. v–x). Cambridge University Press.
Cowell, E. B. (Ed.). (1901). Other sow (Samudda-vānija-jātaka). In W. H. D. Rouse (Trans.), *The Jātaka, or Stories of the Buddha's Former Births* (Vol. 4, pp. 98–104). Cambridge University Press.
Desjardins, J. R. (2013). *Environmental ethics: An introduction to environmental philosophy*. Wadsworth Cengage Learning.
Finnigan, B. (2017). Buddhism and animal ethics. *Philosophy Compass, 12*(7), e12424.
Garrard, G. (2017). Towards an unprecedented ecocritical pedagogy. In *Teaching literature* (pp. 189–207). Springer.
Gombrich, R. F. (2006). *Theravada Buddhism: A social history from ancient Benares to modern Colombo*. Routledge.
Haraway, D. (2008). *When species meet*. University of Minnesota Press.
Harremoes, P., Gee, D., MacGarvin, M., Stirling, A., Keys, J., Wynne, B., & Vaz, S. G. (2013). *The precautionary principle in the 20th century: Late lessons from early warnings*. Routledge.
Harris, I. (2000). Buddhism and ecology. In D. Keown (Ed.), *Contemporary Buddhist ethics* (pp. 113–131). Curzon Press.
James, S. P. (2013). Buddhism and environmental ethics. In S. M. Emmanuel (Ed.), *A companion to Buddhist philosophy* (pp. 599–612). John Wiley & Sons.
Jory, P. (2002). Thai and Western Buddhist scholarship in the age of colonialism: King Chulalongkorn redefines the Jatakas. *The Journal of Asian Studies, 61*(3), 891–918.
Jory, P. (2016). *Thailand's theory of monarchy: The Vessantara Jataka and the idea of the perfect man*. State University of New York Press.
Kahn, R., & Humes, B. (2009). Marching out from Ultima Thule: Critical counterstories of emancipatory educators working at the intersection of human rights, animal rights, and planetary sustainability. *Canadian Journal of Environmental Education (CJEE), 14*, 179–195.
Keiser, D. L., & Adarkar, A. (2015). Buddhas still in classrooms: Where is the mustard seed? *Reflective Practice, 16*(6), 836–848.
King, R. (2013). *Orientalism and religion: Post-colonial theory, India and "The Mystic East"*. Routledge.
Latour, B. (2005). *Reassembling the social: An introduction to actor-network-theory*. Oxford University Press.
Levinas, E. (2011). *Totality and infinity: An essay on exteriority*. Duquesne University Press.
Lewis, T., & Kahn, R. (2010). *Education out of bounds: Reimagining cultural studies for a posthuman age*. Springer.
Lovejoy, A. O. (1936). *The great chain of being: A study of the history of an idea*. Harvard University Press.
Murphy, P. D. (1991). Coyote midwife in the classroom: Introducing literature with feminist dialogics. In J. M. Cahalan & D. B. Downing (Eds.), *Practicing theory in introductory college literature courses* (pp. 161–176). National Council of Teachers of English.
Nixon, R. (2011). *Slow violence and the environmentalism of the poor*. Harvard University Press.
Nussbaum, M. C. (1997). *Cultivating humanity: A classical defense of reform in liberal education*. Harvard University Press.

Oldenberg, H. (1912). The prose-and-verse type of narrative and the jātakas. *Journal of the Pali Text Society, 6*(1908–12), 19–50.
Pierce, D. C. (1969). The middle way of the jātaka tales. *Journal of American Folklore*, 245–254.
Puig de la Bellacasa, M. (2017). *Matters of care: Speculative ethics in more than human worlds*. University of Minnesota Press.
Rhys-Davis, T. W. (1917). *Buddhist India*. T. Fisher Unwin.
Ribó, I. (2012). *Habitat: The ecopolitical nation*. Mycelia Books.
Ribó, I. (2013). Worlds and words: Of bats, ticks, and apes. *Environmental Philosophy, 10*(2), 97–112.
Ribó, I. (2017). Ecocriticism, hermeneutics, and the vanishing elephants of Thailand. In J. C. Ryan (Ed.), *Southeast Asian ecocriticism: Theories, practices, prospects* (pp. 37–60). Lexington Books.
Ribó, I. (2022). Poetics of cohabitation: An ecosemiotic theory of oral poiesis. *Poetics Today, 43*(3), 549–581. https://doi.org/10.1215/03335372-9780431
Ribó, I. (2023a). *Assembling more-than-human sustain-abilities in the ecocritical classroom: A posthumanist approach to the literary curriculum*. Unpublished Master's thesis. University of Gothenburg.
Ribó, I. (2023b). From global citizenship to Anthropocene denizenship: The challenge to education for sustainable development. *Critical Studies in Education, 65*(1), *75–92*. https://doi.org/10.1080/17508487.2023.2222762
Sahni, P. (2008). *Environmental ethics in Buddhism: A virtues approach*. Routledge.
Sahni, P. (2012). Environmental ethics in the Buddhist jātaka stories. *Religions, 2012*(1), 15.
Said, E. W. (1979). *Orientalism*. Vintage Books.
Schmithausen, L. (1997). The early Buddhist tradition and ecological ethics. *Journal of Buddhist Ethics, 4*(1), 1–74.
Sheravanichkul, A. (2008). Self-sacrifice of the Bodhisatta in the Paññāsa Jātaka. *Religion Compass, 2*(5), 769–787.
Singer, P. (1981). *The expanding circle: Ethics and sociobiology*. Farrar, Straus Giroux.
Singh, A. (2015). Buddhist enviromentalism: Narratives from the jātakas. *Journal of the Royal Asiatic Society of Sri Lanka*, 59–79.
Singhsa, K. (2011). Ethics instruction during the period of Thai educational reform. *Prajñā Vihāra: Journal of Philosophy and Religion, 12*(2).
Snaza, N. (2013). Bewildering education. *Journal of Curriculum and Pedagogy, 10*(1), 38–54.
Snaza, N. (2019). *Animate literacies: Literature, affect, and the politics of humanism*. Duke University Press.
Snaza, N., Appelbaum, P., Bayne, S., Carlson, D., Morris, M., Rotas, N., Sandlin, J., Wallin, J., & Weaver, J. A. (2014). Toward a posthuman education. *Journal of Curriculum Theorizing, 30*(2), 39.
Snaza, N., & Weaver, J. A. (2015). *Posthumanism and educational research*. Routledge.
Tassoni, J. P. (1998). Deep response: An ecofeminist, dialogical approach to introductory literature classrooms. In G. Gaard & P. D. Murphy (Eds.), *Ecofeminist literary criticism: Theory, interpretation, pedagogy* (pp. 204–223). University of Illinois Press.
Taylor, A. (2017). Beyond stewardship: Common world pedagogies for the Anthropocene. *Environmental Education Research, 23*(10), 1448–1461.
Thathong, K. (2012). A spiritual dimension and environmental education: Buddhism and environmental crisis. *Procedia-Social and Behavioral Sciences, 46*, 5063–5068.
Todd, S. (2015). *Toward an imperfect education: Facing humanity, rethinking cosmopolitanism*. Routledge.
Viveiros de Castro, E. (1998). Cosmological deixis and Amerindian perspectivism. *Journal of the Royal Anthropological Institute, 4*(3), 469–488. https://doi.org/10.2307/3034157
von Hinüber, O. (1998). *Entstehung und Aufbau der Jātaka-Sammlung*. Akademie der Wissenschaften und der Literatur; F. Steiner.

Waldau, P. (2000). Buddhism and animal rights. In D. Keown (Ed.), *Contemporary Buddhist ethics* (pp. 83–111). Curzon Press.
Wals, A. (2012). Learning our way out of unsustainability: The role of environmental education. In S. D. Clayton (Ed.), *The Oxford handbook of environmental and conservation psychology*. Oxford University Press.
Wals, A., Geerling-Eijff, F., Hubeek, F., van der Kroon, S., & Vader, J. (2008). All mixed up? Instrumental and emancipatory learning toward a more sustainable world: Considerations for EE policymakers. *Applied Environmental Education and Communication, 7*(3), 55–65.
Weik von Mossner, A. (2017). *Affective ecologies: Empathy, emotion, and environmental narrative*. The Ohio State University Press.
White, L., Jr. (1967). The historical roots of our ecologic crisis. *Science, 155*(3767), 1203–1207.

Ignasi Ribó (PhD, Universitat de Barcelona/University of Sussex) is an Associate Professor of Comparative Literature at the School of Liberal Arts, Mae Fah Luang University (Chiang Rai, Thailand). His research deals with issues in the Environmental Humanities, bridging across the interdisciplinary fields of Ecosemiotics, Ecocriticism, Education for Sustainability, and Human Ecology. He is the author of *Habitat: The Ecopolitical Nation* (2012) and a contributor to the edited collection *Southeast Asian Ecocriticism: Theories, Practices, Prospects* (2017). He has also published the textbook *Prose Fiction: An Introduction to the Semiotics of Narrative* (2019), as well as five novels and numerous academic essays in international journals. He is a member of the Executive Council of the Association for the Study of Literature and Environment in ASEAN (ASLE/ASEAN).

Chapter 6
The Multicultural Almanac as a Pedagogical Tool in Religious Education

Linda Jonsson

6.1 Introduction

Calendar and almanac are terms that have long historical roots and can be seen as synonymous with the measurement of time. A calendar provides information on, for example, the months, weeks and days of the year, but also expresses astronomical phenomena such as the change of seasons, sunrise and sunset, and the different phases of the moon. The term almanac is customarily used when information on for example traditions, holidays, memorial days and recurring events of general interest are added to the calendar. An almanac can also express recurring religious holidays, traditions, popular celebrations and ceremonies with historical roots far back in time. The first printed almanac, from a European perspective, came out in Germany in 1448 and the first Swedish almanacs were direct translations of the German original. The Swedish almanacs from the start of the seventeenth century featured health advice, the weather, divinations and moral teachings of various kinds. The almanac gave notice of what had happened but also of what was going to happen. An appendix, Prognostikon, came with the almanac, which consisted of predictions of war, the year's crops, the reader's health, illnesses and death, all of this at times specified in the form of horoscopes (Hansson, 1986).

Another type of almanac was the Farmer's Almanac [Bondepraktikan], which was also based on a German almanac from the start of the sixteenth century, and became popular in the Nordic countries. Initially the Farmer's Almanac was a handbook for farmers and included prognoses and predictions about the weather, but later came to include health advice, descriptions of national holidays and traditions

L. Jonsson (✉)
School of Education, Culture and Communication, Mälardalen University, Västerås, Sweden
e-mail: linda.jonsson@mdu.se

S. Windsor, O. Franck (eds.), *Intersections of Religion, Education, and a Sustainable World*, https://doi.org/10.1007/978-3-031-81809-7_6

of both Christian and popular origin. Holidays such as Christmas, Easter and Midsummer's Day were described both as national holidays with their Christian content and as popular celebrations connected to the seasons and older folk beliefs. In addition to the national holidays, Swedish festivals that celebrate various aspects of Swedish traditions were listed. The almanacs were popular in everyone's home and were printed in large editions for the time, thus making them an important channel for the dissemination of facts and knowledge of general interest (Hansson, 1986). To this day, the Farmer's Almanac is published with recurring themes such as holidays, weather forecasts, health and farming.

6.2 The Multicultural Almanac

An almanac with an international starting point has been published in Sweden since the 1990s. Named the multicultural almanac, (Mångkulturella almanackan, 1) it is a wall calendar intended for the schools' classrooms, workplaces and the home. This almanac stems from a time before the internet was available in schools, in classrooms and for teaching. The Swedish Immigrant Institute and Museum, (Sveriges Invandrarinstitut och Museum), which later became The Multicultural Centre (Mångkulturellt centrum -MC), planted the seed for an almanac taking an international, outward-looking perspective, as schools were reaching out to them with questions about when Ramadan occurs for example. The Institute then contacted SIOS, Samarbetsorgan för invandrarorganisationer i Sverige (the cooperation body for immigrant organisations in Sweden), which still exists today, with the following appeal:

> Many holidays that are not generally known are celebrated in Sweden, and we would therefore like to help draw attention to them. We would like to have your help to list the days that are of particular importance to the members of your association in particular. The days we would like to include in the calendar are
>
> - religious holidays
> - national holidays
> - national celebrations or memorial days of a different kind (https://mkalmanackan.se/om-oss/)

The multicultural almanac is now published annually, announcing, the major religions' regularly recurring holidays and commemorations, folk festivals and celebrations, UN days and the days of Sweden's national minorities among other things. The multicultural almanac also has a website that is free to use as a starting point for the school's teaching. Under the heading of Pedagogical pearls, the website regularly publishes materials with suggestions and exercises for teaching, from preschool to adult education. Every month Pedagogical pearls are published on the website in the form of a two-sided, printable PDF that is free to use in school teaching. This is the multicultural almanac that is used as an example in this chapter.

6.3 Religious Education in a Swedish Context

Over a period spanning 200 years, the history of the subject of religious education in Sweden and the Nordic countries has undergone major changes. Two hundred years ago the school's sole task was to foster good Christian citizens. The Lutheran faith was the point of departure and the aim of the teaching focused on devoutness in the sense that every pupil was expected to fear and honour God. In this early elementary school, religion was the most prestigious school subject (Algotsson, 1975; Lundgren, 1979).

What was the situation like, then, for the teachers who were serving in elementary school in the middle of the nineteenth century onwards? The prevailing form of teaching was strongly influenced by the technique the clergymen practised at the parish catechetical meetings in people's homes, based on Luther's Little Catechism. Based on these recurring parish catechetical meetings, a pattern with a question–answer model was developed, where the clergyman asked questions and then made sure that the members of the parish had the right answers. Catechisation, as this model came to be named, was practised diligently in schools. It was not until some time into the twentieth century that attention was directed to the pupils in the classroom, who they were and what experiences they had. Along with the curriculum that came out in 1919 followed recommendations for a new pedagogy with the pupil at the centre. The new curriculum also included an outward-looking perspective that considered other religions than Christianity. However, it took a long time, until the school reforms of the 1960s, before the school subject of education in Christianity (*kristendomskunskap*) was replaced by religious education (*religionskunskap*). Both the teaching and the subject have thus changed from denominational Christianity in a school operating in a fairly uniform society to teaching in a school system and society characterised by diversity (Jonsson, 2016). Today's classrooms can be likened to a melting pot where different cultures, religions and traditions mix and meet. One of the perhaps most important tasks of schools today is to give pupils the skills they need to face both a globalised world and an increasingly multifaceted local community with a spectrum of worldviews, outlooks and ideas.

In the Swedish curriculum for preschool class to year 3 (F–3) (Skolverket, 2022), the introductory text on the school's values states that the school is to portray and convey the values and rights that are expressed in the UN Convention on the Rights of the Child. It also highlights the role that the school has for the pupils' confidence and self-esteem. This is expressed in statements declaring that intolerance and oppression must be prevented and encountered with knowledge and active measures, and that everyone in school is to take a democratic and norm-conscious approach as a starting point for example (Skolverket, 2022, pp. 5–10).

Turning to the syllabus for religious education, it is specified that the social studies subjects have the responsibility to teach an understanding of people's living conditions. In other words, it is about mutual understanding between people so as to create a good life for oneself and others in a complex world. For religious education (RE), it is especially emphasised that the subject is warranted by today's multicultural society where ‘knowledge about religions and other outlooks on life [are] important for creating mutual understanding between people’ (Skolverket, 2022, p. 188). In the statement of purpose for RE it is stated that the pupils are to be given the opportunity to develop ‘knowledge on religions and other outlooks on life as well as on different interpretations and varying practices within them’ (Skolverket, 2022, p. 188). The overarching aim and purpose of the teaching in religion is that everyone, regardless of background, can meet on the same terms. An explicit aim of RE is to reach a greater understanding of the richness of diversity.

In the syllabus, RE has some shared content with the other social science subjects, where the core content is linked together under four headings:

- Living together
- Living in the neighbourhood
- Living in the world
- Exploring reality

The headings thereby indicate that education in social studies to a large extent is subject integrated, even though the subject of religion is presented with specific content. The central aims of RE clearly feature ethics and studying the religions that can be found in the nearby area.

6.3.1 The Didactic ‘Who’ Question with Today’s Classroom in Focus

To teach, content must be defined, chosen and made teachable. When the teacher makes a didactic plan, the starting point is determined based on the didactic questions of ‘what’, ‘how’, ‘who’ and ‘why’ to determine what is to be taught (content) and in what way (transmission). The question of ‘why’ is about the specifications in the curriculum concerning the school’s organisation, mission and content and the directives aimed at the school’s subjects (Jank & Meyer, 1997). The relation between the teacher, the content and the one who takes part in the teaching, namely the pupil, is usually called the didactic triangle, which illustrates the interplay between the three in the teaching. In addition to the teachers choosing content with their subject knowledge, this content is also adapted and reworked to be made teachable. It is a question of how the content is transmitted to the pupils and with what methods. Yet another factor to take into consideration is the relation teachers have to their students, as the teaching constitutes a structured togetherness around the pupils’ individual and joint learning processes.

The didactic ‘who’ question puts the focus on who the pupils in the classroom are, as the goal of the teaching is to offer something meaningful to every pupil. The composition of classrooms, in terms of the pupils in Swedish school today looks much different today than to earlier when Christianity was part of a Swedish national identity. Religious diversity and pluralism are an indisputable part of contemporary

Sweden and leave their mark on today's classroom. Migration to Sweden has increased in recent decades and today around 20% of people living in Sweden are born abroad. Religious traditions from the rest of Europe, the Middle East, Asia and Africa are establishing organisations across the country and Sweden now ranks first among European countries in terms of religious diversity (Willander, 2019). Although Sweden today is an ideologically secular country, religion is still integrated in social practices and everyday life in many ways, influencing how we think about for example traditions and holidays as content in teaching. The target group of the teaching is not an anonymous and homogeneous group of pupils, but rather children with their own voices and will, and who bring with them various religious backgrounds and ideas. The 'who' question thus points towards making visible the pupils that populate the classroom in terms of their knowledge, interests and experiences, and letting them express themselves as unique individuals with their own voices.

The didactic 'who' question also involves how the teacher chooses to organise the classroom for the pupils' benefit and how the teaching can make room for existential questions concerning man's place in the world. This is connected to questions concerning what makes us human in a social context. The didactic choices a teacher makes by interpreting and converting the syllabus in religion into practice will be decisive for the content that in fact is in focus. School is the place where many young people encounter religion and faith in a secular context for the first time, and that is why a particular responsibility rests on the teacher when it comes to rendering nuanced representations of religious traditions and holidays. Several of the holidays that are observed within the framework of the school's teaching have Christian roots, such as Lucia, Christmas and Easter. In this context, Ann af Burén (2016) points to how, considering the school year, there are plenty of occasions to discuss what we, teachers, pupils as well as society, understand as religion in Sweden today in connection to holidays and traditions. Is the Lucia procession a religious holiday, a specifically Christian holiday or a recurring tradition? Is the fact that the children have Easter and Christmas holidays, or that we measure our time starting with the birth of Christ, a sign that we live in a Christian culture (af Burén, 2016)? It is also about including elements that have not sprung from a traditional heritage with Christian roots, as Tünde Puskás and Anita Andersson (2017, 2020) highlight in a study focusing on preschool practices. They problematise to what extent what is celebrated in preschools can be described as cultural heritage and something that is associated with religions. The challenge then lies in whether what is said and done is in agreement with the requirements of non-denomenationalism described in the preschool curriculum. They find that as long as the teaching of traditions does not include any denominational elements such as prayers, blessings or professions of faith, preschools can convey religious traditions.

To sum up, the 'who' question puts the focus on those in the classroom, right there in the middle of the action in a given situation. With that in mind, the multicultural almanac as a starting point and source of knowledge can be a springboard to creating a more democratic and inclusive presence in the classroom.

6.3.2 An Intercultural Pedagogical Approach

With the report Different origins – community in Sweden. Education for linguistic and cultural diversity [Olika ursprung – gemenskap i Sverige. Utbildning för språklig och kulturell mångfald] (SOU, 1983:57), the terms multicultural and intercultural were put under scrutiny. The report states that teaching should be adaptable to the presence of many languages and cultures in the classroom. No clear definition of intercultural teaching is given, but what is established is the notion that intercultural teaching concerns every child and every adult both in school and in the surrounding society. In the report a distinction is made between the terms multicultural and intercultural based on the Council of Europe's definitions, which state that multicultural is used in the sense of a state of being, situation or position. In other words, it is about seeing school as a place where there are pupils who represent different cultures, ethnicities or nationalities. The term intercultural, in contrast, indicates an action, something that happens and has to do with movements between individuals:

> If we, for example, in a multicultural school have the ambition to develop a new way of relating to one another, one that makes demands on everyone in school as a result of the multicultural situation, then this requires intercultural teaching and an intercultural approach as well as an intercultural leadership of the school's leaders. (Lorentz & Bergstedt, 2016, p. 16)

This points to a way of relating to one another in teaching based on an intercultural approach. Lorentz and Bergstedt (2016) maintain that it is about interaction and reciprocity between people that also crosses boundaries. A precondition and point of departure for an intercultural perspective and approach in teaching is an acknowledgement that a multicultural teaching environment consists of individuals with different cultural perceptions and values.

In this context, let us for a moment lift up RE to an intercultural level where the different religions and outlooks on life are not just seen as bounded, separate units, but are seen in a larger and broadened perspective (Heimbrock, 2007; Lorentz & Bergstedt, 2016; Lahdenperä, 2018). What does it mean to acquire intercultural pedagogical competence to teach RE? Pirjo Lahdenperä (2018) maintains that intercultural competence can be described in terms of how a person perceives and interprets multicultural differences. These differences can in turn be described in many different ways and from different perspectives. The social constructionist approach to looking at culture concentrates on the construction of meaning and how this meaning is created and applied in a cultural context, for example in a classroom. Lahdenperä (2018) suggests that there are some starting points for describing cultural and ethnocultural elements in teaching:

- Cultural artefacts – what different expressions can be distinguished when it comes to food, music, clothing for example?
- Repeated patterns of behaviour – what different traditions, holidays and ceremonies do we carry with us from our experiences?

- Collective religious beliefs and belief systems – what are the values, norms and perceptions of what is right and wrong?
- Our thinking and feelings are culturally bound. What different ideas, metaphors and emotional expressions do we carry with us?
- The way of communicating with and relating to one's surroundings – what do family relations and relations between the genders look like?
- One's own self-concept – how is one's own self-image constructed?

The above points of departure form the basis for acquiring an interculturally competent approach in teaching. This expands the possibilities of planning RE with a broader content and perspective, taking into consideration who we are and what comes with us into the classroom in the form of ways of thinking, experiences and traditions.

6.4 RE and the Lifeworld

Before looking at teaching in RE and its didactic implications with a starting point in the multicultural almanac, there is reason to pause for reflection on religion in relation to the environment both locally and globally. The German scholar of religion Hans-Günter Heimbrock (2005) has formulated a didactic approach for religion that starts from the pupils' lifeworld, where he formulates questions about how young people orient themselves in a constantly changing world and how their questions about life can be attended to in the best possible way. The approach for school and teaching formulated by Heimbrock can be described in two ways:

- Each country has its own specific religious educational development and its own specific way of designing its teaching in school. European integration also becomes particularly noticeable in this area. Reflections on the individual teachers' contexts are highlighted. The term that is emphasised in this approach is lifeworld-oriented religious education.
- Theoretically, the approach builds on impulses from philosophical phenomenology and emphasises a view of religion that to a very high degree has to do with existence as a whole. This opens up new opportunities to approach religious phenomena in a cultural context.

The term lifeworld cannot easily be defined, but Heimbrock formulates it as:

> Lifeworld is to be understood as the given, familiar, inevitable and concrete reality of everyday life. In that sense, every person has their own lifeworld. The phenomenological term lifeworld captures the cultural perspective. 'The meaning of the small world'. (Heimbrock, 2005, p. 12)

Based on these perspectives, Heimbrock expands on some thoughts on what religion can be thought about from a lifeworld perspective. First of all, good teaching opens up new contexts. Religion becomes clearer in the diversity and forms of expression of everyday culture. It also offers the opportunity to look and reflect with everyday life and the lifeworld as the starting point, and religion with its forms of expression becomes possible to describe contextually. Heimbrock's (2005) interest when it comes to content takes its starting point in a religious pedagogy designed in close connection with today's Western European culture. A dialogue with different approaches challenges earlier viewpoints and experiences and thereby opens up new opportunities for continued development of the nearby work in the classroom as well as different ways of relating to the surrounding society at large.

Heimbrock (2005) maintains that based on the fact that religion already is present in everyday life and 'in reality's ever-present background' (p. 29), it's a matter of focusing on people's encounters with this reality and also with what is foreign. The essential thing is to create an openness. Based on context, religion emerges as stronger in its diversity and forms of expression in everyday life. Heimbrock argues that understanding the didactic triangle is not enough, rather consideration of the context is critical when teaching religion. A broadened concept of religion is also necessary for interpreting modern forms of expression of religious practice contextually, in both historical and present-day contexts. Connected to this is a need for a dual conceptualisation of religion (Heimbrock, 2005):

1. Religion consists, among other things, of denominational systems, collective traditions and worldviews.
2. Religion is also more than what is defined in the point above. Religion also includes a process that manifests itself in that people develop and shape their personality.

This broadened concept of religion leads to what Heimbrock (2005, 2007) calls lived religion. This term makes it possible to look at religion in a way that has to do with one's whole existence, and with a broadened concept of religion it is possible to examine expressions of religious practices contextually, both in a historical and in a contemporary context.

The Multicultural Almanac's Contents – Three Examples

The multicultural almanac has a traditional day-by-day format, announcing both national and international events. Every edition of the multicultural almanac has a theme that runs through all months and days. Based on the current theme, stories from different traditions, religions and international events are woven together. In Fig. 6.1, examples from the three editions from the years 2020, 2022 and 2023 are illustrated.

The 2020 edition of the multicultural almanac is introduced with the heading 'All tradition is change'. This edition has food as its theme throughout, in the sense of food being where different traditions meet. Here you find stories and interviews about meal traditions and celebrations, and also recipes and cultivation tips from a sustainability perspective. Throughout the almanac attention is drawn to, among other things, 30 new year's holidays from different religious and national perspectives, along with explanatory texts. First is the Jewish new year Rosh Hashana, where it is customary to eat a piece of apple dipped into honey, with the meaning of wishing everyone a sweet and blessed year. Thereafter, Vietnamese new year is

Fig. 6.1 The cover image from the 2020, 2022 and 2023 almanacs (https://mkalmanackan.se/)

observed, where the celebrations include eating different types of small pancakes. The Vietnamese celebration's food culture is then connected to different types of pancakes in different cultures, traditions and religions. Here you find descriptions and stories about Swedish food traditions with oven-baked pancakes and small pancakes, the Åland tradition with Åland pancakes, Eritrean and Somali sourdough pancakes, injera and laxoox as well as canjeero, respectively. The theme is finally summed up with a direct question: What about you?

Month after month the changing seasons are addressed, featuring cultivation tips. For Easter, which is highlighted as both a Jewish and a Christian tradition, there is a description of how one can grow Easter grass: 'After a few days, the grass will grow and you can place small chicklings, eggs and other Easter decorations in it'. In the month of July a salad dish from Asia is explained in both Swedish and Arabic in the almanac. An explanation of the dish is offered, its origins, how it is grown in Sweden and how the dish is adapted to be sustainable:

> Fatoush is a salad dish from south-west Asia. We got this recipe in Arabic from a cultivator at the allotment association Lilla Rinkeby in Stockholm. Fatoush is a good way of making use of some old bread, pita or another sort. Fry or bake it in the oven until it gets a nice colour, break it into pieces and garnish the salad with it.

The multicultural almanac for the year 2022 has a nature theme throughout the year's months and days. Trees are in focus:

> Trees, our large friends. Trees were on earth when humans started to wander about. They have followed us over thousands of years. [...] Trees give us fruit and nuts, they give us timber to do woodwork and build things. They can be used for industrially produced paper pulp and they can be used to make a fire. Humans learned to make fire half a million years ago and the power over fire has been at the centre of human culture. Large bonfires or smaller fires mark the shifting seasons. We will tell you about spring's Festival of Fire and summer's Midsummer fires. In September we write about the fires at the Orthodox celebration of the Holy Cross.

Apart from telling us about different traditions and holidays, throughout the almanac there are features about how humans influence nature, with a call for reflection. Although a gloomy image is given of how humans damage nature, the almanac is permeated with a sense of hope for the future, such as:

> Too much is burning now. Forests are burned down as biofuel and forest fires rage in the summer. For 2022, a burning current almanac, with memories and with hope for the future. Let the trees grow. And people live.

The stories about trees are also related to different religions. The Bodhi tree is talked about in relation to Buddhism, the Sidr tree is at focus in the story about the prophet Muhammad's ascension, Lailat al Miraj, the little myrtle tree is described in connection with the Feast of St John the Baptist, where every participant in Mandaean tradition is given a twig of myrtle when baptised. We are told about the Jewish Feast of Tabernacles and how this was celebrated for the first time in Sweden in 1685. The trees are also linked to all of creation where the major world religions make clear appeals to how man has a mission to respect nature. In this context, the multicultural almanac emphasises that interreligious environmental networks and organisations have emerged around the world.

Movement is the theme for the year 2023, where we are told about holidays and days that are based on some form of actual movement such as pilgrimages and ascensions, but also stories about escape and migration. The introduction explains movable religious feasts and how these are connected to the phases of the moon. Examples include why it is that Ramadan is moved 'back' in time in the Gregorian calendar every year, how Christian Easter is influenced by the moon, how Jewish holidays are based on both the moon and the sun, as well as how the Chinese, Korean and Vietnamese new year take place on the second new moon after the winter solstice.

Popular movements are also given attention, looking back to the nineteenth century and onwards when people organised themselves around current societal issues. The multicultural almanac accounts for some popular movements that still exist, but also shows new ones that have been formed and how people organise themselves around issues that are important to them. Among other things, we learn about the Travelling People's day, which is one of the newer traditions that has been added:

> The first written note about Romani presence in Sweden is from 29 September 1512, when it was written in the records of the City of Stockholm that a group had come to town. This is why the Travelling People's day is celebrated on 29 September.

Animals also have their place in the almanac. We learn about how the animal rights movement emerged based on how Saint Francis was declared patron saint of animals in 1939 and patron saint of ecology in 1980. Saint Francis' day is 4 October, which also is the Day of the Animal Rights Movement and World Animal Day in the almanac.

The multicultural almanac in 2023 also addresses how movements can be going in the wrong direction. In November a background as to why the date of 9 November has a place in the almanac is given, considering how Jewish-owned shops had their windows crossed over all of Germany during the Crystal Night on that date in 1938. Taking this Pogrom as a starting point, we are called upon to look at present times:

> Today we can unfortunately see currents that seem to remind one of the 1930s. The populist nationalism that is gaining ever more ground is characterised by hatred of 'foreigners' and an insularism that can threaten democracy and risk people's safety. The persecution of and threats against religious minorities and those who break the majority society's norms are daily occurrences.

In this context we are also called upon to talk about how all of us together can work on distancing ourselves from the threats we face in various ways: 'To stand up for one another every day, for everyone's human dignity, THAT is something sparkling and nice'.

Last in the 2023 edition of the multicultural almanac, International Migrants Day is observed and explained. Here a historical background is given from the nineteenth century and the start of the twentieth century, when many Swedes left their homes and became immigrants in the North America. The reasons for emigration that are stated are above all poverty, lack of arable land, religious persecution, unemployment and escape from conscription. It is stated that today around 3.6% of the world's population is fleeing from poverty and conflicts, with the date of 18 December in the almanac.

6.5 Pedagogical Implications for Teaching of Religion and Sustainability

The Swedish Public Health Agency (Folkhälsomyndigheten, 2022) as well as The 2030 Agenda defines social sustainability as an equal and equitable society where people live a good life with trust and confidence in each other. The 2030 Agenda is a global initiative that highlights the UN's 17 Sustainable Development Goals. The purpose of these goals is, among other things, to achieve an equal society regardless of where you live. One of the goals emphasises that education is a prerequisite for ensuring an inclusive and equal society (Regeringskansliet, n.d.).

When it comes to the teaching of RE for the younger ages and the aim and content of the syllabus, with the headings of Living together, Living in the neighbourhood, Living in the world and Exploring reality, the multicultural almanac holds value for teachers. The syllabus requirements suggest that the almanac is warranted by today's multicultural society where 'knowledge on religion and other outlooks on life [are] important for creating mutual understanding between people' (Skolverket, 2022, p. 188). This puts the spotlight on taking responsibility and understanding people's lifeworlds. In other words, it is about mutual understanding between people in a complex world, the local community and the world at large, to create a good life for oneself and others. Based on the writings in the syllabus and the multicultural almanac as a pedagogical tool, some didactic implications are identified below to create a teaching context in RE that is socially sustainable and adapted to the societal context.

6.5.1 *The Didactic 'Who' Question Is Negotiated into a 'We' in the Classroom*

With the multicultural almanac as content in RE, the didactic 'who' question is put into focus, and here we find propositions that make it possible to construct a multifaceted 'we' that embraces diversity and takes into account our different backgrounds and who we are in the classroom. Society as well as school and traditions are processes on the move that perpetually change over time. Faith, traditions and cultural heritage come with every pupil in one way or another. When it comes to the school's teaching on different holidays in the different world religions, apart from creating a context and an understanding, it can also be a subject area filled with passion as holidays often are associated with joy and celebration. The multicultural almanac accounts for holidays, ceremonies and food associated with specific events. By taking hold of such elements in terms of cultural artefacts (Lahdenperä, 2018), it becomes possible to take an intercultural approach to teaching RE with intercultural competence of action. In this way, the multicultural almanac can be the link that ties together the writings of the syllabus with who we, the ones who are populating the classroom, are at a given point in time, and what the stories will be that are negotiated based on this content. Lahdenperä also underlines that thinking and feelings are culturally bound and form the basis for the ideas, metaphors and emotional expressions we carry with us into the classroom. This in turn touches on how one's own concept of self and one's own self-image are constructed based on such content.

6.6 Lifeworld-Oriented RE

Relating to the headings in the syllabus for RE calls for an approach that binds the teacher to choose content and methods that allude to creating a good life for oneself, for others in the local community but also with a global outlook. The multicultural almanac's content with all sorts of events in the near and far, capture the pupils' own lifeworld in the diversity and forms of expression of everyday culture; it is about one's whole existence, where traditions and religions find their place in the classroom's specific cultural context. Based on Heimbrock's argument focusing on the pupil's lifeworld and with a broadened concept of religion, the multicultural almanac offers didactic implications for shedding light on collective traditions and worldviews in teaching. Here, there is room for everyday culture's proximity, tangibility and expressions from a perspective close to the pupils.

A lifeworld-oriented RE also opens up for perspectives that feature nature. An almanac's trademark in general terms is to inform about, for example, the months, weeks and days of the year, but it also expresses astronomical phenomena such as shifting seasons, sunrise and sunset as well as the phases of the moon. The multicultural almanac (regardless of edition) additionally includes content that takes nature into consideration. The changing seasons are illustrated, as are what is cultivated and the different ways in which people relate to this. It is about people morally and ethically taking responsibility for nature, plants and animals, which connects with the writings in the syllabus where the pupil is to be able to 'analyse and take a stand in ethical and moral questions' and that the teaching should 'contribute to the pupils' possibilities to develop a personal attitude to life and a preparedness to act responsibly in relation to themselves and their surroundings' (Skolverket, 2022, p. 188). The multicultural almanac as a pedagogical tool offers a sustainable approach to diversity, nature and culture for teaching on religion and sustainability.

References

af Burén, A. (2016). Bilder av kristendom i det samtida svenska samhället – med blick mot historien [Images of Christianity in contemporary Swedish society – With an outlook on history]. In O. Franck, C. Osbeck, & K. von Brömssen (Eds.), *Religioner, livsåskådningar och etik* [Religions, outlooks on life, and ethics]. Gleerups.

Algotsson, K.-G. (1975). *Från Katekestvång till religionsfrihet. Debatten om religionsundervisningen i skolan under 1900-talet* [From forced catechism to freedom of religion. The debate on teaching religion in school during the twentieth century]. Rabén & Sjögren.

Folkhälsomyndigheten. (2022). https://www.folkhalsomyndigheten.se/motesplats-social-hallbarhet/social-hallbarhet/ (Last accessed 20 July 2023).

Hansson, S. (1986). 1600-talsalmanackan – Allemansbiblioteket [The almanac of the seventeenth century – The public library]. *Lychnos, 1986*, 29–53.

Heimbrock, H.-G. (2005). *Livsfrågor – Religion – Livsvärld. Bidrag till kontextuell religionsdidaktik ur ett tyskt perspektiv* [Life questions – Religion – Lifeworld. Contributions to contextual didactics of religion from a German perspective]. Författaren och RPI – Arbetsgemenskapen för religionspedagogik.

Heimbrock, H.-G. (2007). Reconstructing lived religion. In H.-G. Heimbrock & C. P. Scholz (Eds.), *Religion: Immediate experience and mediacy of research. Interdisciplinary studies in the objectives, concepts and methodology of empirical research in religion*. Vandenhoeck & Ruprecht.

Jank, W., & Meyer, H. (1997). Nyttan av kunskaper i didaktisk teori [The usefulness of knowledge of didactic theory]. In M. Uljens (Ed.), *Didaktik: teori, reflektion och praktik* [Didactics: Theory, reflection and practice] (s. 17–34). Studentlitteratur.

Jonsson, L. (2016). *Mellan tradition och förnyelse – Utmaningar i religionslläraruppdraget* [Between tradition and renewal – Challenges in the religious studies teacher assignment]. Diss. Mälardalens högskola, Västerås.

Lahdenperä, P. (Ed.). (2018). Inledning [Introduction]. In P. Lahdenperä (Ed.), *Den interkulturella förskolan: mål och arbetssätt* [Intercultural preschool: Aims and ways of working]. Liber.

Lorentz, H., & Bergstedt, B. (2016). Interkulturella perspektiv [Intercultural perspectives]. In H. Lorentz & B. Bergstedt (Eds.), *Interkulturella perspektiv: pedagogik i mångkulturella lärandemiljöer* [Intercultural perspectives: Pedagogy in multicultural learning environments]. Studentlitteratur.

Lundgren, U. P. (1979). *Att organisera omvärlden. En introduktion till läroplansteori* [Organising the world around us. An introduction to curriculum theory]. Liber Förlag.

Puskás, T., & Andersson, A. (2017). "Why do we celebrate …?" filling traditions with meaning in an ethnically diverse Swedish preschool. *International Journal of Early Childhood, 49*, 21–37. https://doi.org/10.1007/s13158-017-0182-8

Puskás, T., & Andersson, A. (eds) (2020). *Traditionernas didaktik: religion och kulturarv i förskolan* [The didactics of tradition: Religion and cultural heritage in preschool] (1st edn). Gleerups.
Regeringskansliet. (n.d.). https://www.regeringen.se/regeringens-politik/globala-malen-och-agenda-2030/agenda-2030-mal-4-god-utbildning-for-alla/ (Last accessed 20 July 2023).
Skolverket. (2022). *Läroplan för grundskolan, förskoleklassen och fritidshemmet, Lgr22* [Curriculum for the compulsory school, preschool class and the leisure-time centre].
SOU 1983:57. *Olika ursprung – gemenskap i Sverige. Utbildning för språklig och kulturell mångfald* [Different origins – Community in Sweden. Education for linguistic and cultural diversity].
Willander, E. (2019). *The Religius landscape of Sweden – Affinity, affiliation and diversity in the 21st century* [Elektronisk resurs]. Myndigheten för stöd till trossamfund – SST.

Links

https://mkalmanackan.se/om-oss/ (Last accessed 19 March 2023).
Mångkulturella almanackan 2020 [The Multicultural Almanac 2020]: https://webshop.mkcentrum.se/varabocker/mangkulturella-almanackan-2020; https://mkalmanackan.se/pedagogiska-parlor/ (Last accessed 18 February 2023).
Mångkulturella almanackan 2022 [The Multicultural Almanac 2022]: https://webshop.mkcentrum.se/varabocker/mangkulturella-almanackan-2022; https://mkalmanackan.se/pedagogiska-parlor/ (Last accessed 18 February 2023).
Mångkulturella almanackan 2023 [The Multicultural Almanac 2023]: https://webshop.mkcentrum.se/varabocker/mangkulturella-almanackan-2023; https://mkalmanackan.se/pedagogiska-parlor/ (Last accessed 18 February 2023).
Mångkulturella almanackan [The Multicultural Almanac]: https://mkalmanackan.se/2018/03/13/i-museets-dolda-vrar-mangkulturella-almanackans-start-1989/ (Last accessed 18 February 2023).

Linda Jonsson is a Senior Lecturer and PhD in didactics in the School of Education, Culture and Communication at Mälardalen University, Sweden. She works in teacher education and her research interest focuses on general didactic and subject didactic issues, especially in relation to democracy, religion and the teaching of religion.

Chapter 7
Religious Education in Türkiye in Terms of Sustainable Environmental Education: An Analysis of Religious Culture and Ethical Knowledge Curriculum and Textbooks

Hasan Meydan, Recep Kaymakcan, and Feyza Karaahmetoğlu

7.1 Introduction

In modern environmental education, for a sustainable environmental management, the most frequently emphasized ethical justification is the moral responsibility of the current generation to future generations. Even though this approach directs people to protect the environment for future generations, the responsibility towards the environment is an indirect responsibility. This is because the environment is placed in a position that is respected not for itself but for the benefit of future generations. However, for religions, all existence deserves existential respect because it is divine in origin. For example, Muslim scholars have developed approaches such as the balance of usefulness, responsibility, virtue and wisdom in the human-environment relationship (Yaran, 2008). It is therefore important to consider these religious approaches in religious education for the development of sustainable environmental awareness. The curriculum[1] of Religious Culture and Ethics Knowledge course, which has been a part of the school compulsory curriculum in Türkiye since 1982, has been changed four times in 1982, 2000, 2006 and 2018. In these curriculums,

---

[1]The Ministry of National Education officially refers to the Religious Culture and Ethics Knowledge curriculum as "teaching programme" (öğretim programı in Turkish), by which it means 'curriculum'. In this article, therefore, we use "curriculum" to refer to RE teaching programme.

H. Meydan (✉)
Religious Education, Sakarya University, Serdivan, Türkiye

R. Kaymakcan
Religious Education, Turkish Ministry of Youth and Sport, Ankara, Türkiye

F. Karaahmetoğlu
Social Sciences Institute, Sakarya University, Serdivan, Türkiye

S. Windsor, O. Franck (eds.), *Intersections of Religion, Education, and a Sustainable World*, https://doi.org/10.1007/978-3-031-81809-7_7

the human-environment relationship has been mentioned in the aims, values, attainment targets and units in different shapes and forms. The subject has been mentioned in topics such as Environmental Cleanliness, Human-Nature Relationship and Environmental Awareness in Religions. Moreover, the terms such as human responsibility to existence, natural sustainability, environmental awareness and balance in production and consumption have been used in the curricula and textbooks.

In this context, firstly, the theoretical dimension pertaining to the concept of sustainable development, its understanding in Türkiye, and Islamic perspective will be examined in line with the current literature. Secondly, it aims to explore how sustainable environmental education has been evaluated in Religious Culture and Ethics Knowledge's curricula and textbooks since 1982, as religious education has a potential to contribute significantly to sustainable environmental awareness. Moreover, it intends to make a suggestion in line with better sustainable environmental education for future curriculum development and preparation of textbooks in schools' religious education.

7.2 Sustainable Development and Türkiye

In recent years, we have frequently heard international calls supported by the knowledge that the world's average temperature has increased by one degree in the last century, because of which drought, famine, unpredictable natural events/disasters, and socio-economic problems have been experienced (United Nations Educational, Scientific and Cultural Organization [UNESCO], 2020, s. 6). If global warming continues at the same pace, it seems certain that the increase will reach 1.5 °C before 2050. The worse scenario is that this increase will reach 2.0 °C if the measures taken by human beings do not work. Limiting global warming to 1.5 °C is expected to substantially reduce the probability of extreme drought, precipitation deficits, and risks associated with water availability (i.e., water stress) in some regions (see The Intergovernmental Panel on Climate Change [IPCC], 2018) It is clear that unprecedented and dramatic changes are taking place and that humankind is directly responsible.

Just as the responsibility of destruction belongs to humans, the development of an understanding that will provide balance is also subject to human ingenuity. It is clear that the devastating impact will continue to widen unless humankind develops a reasonable balance between the use and conservation of natural resources (see UNESCO, 2020). Currently, the most comprehensive understanding developed by mankind is embodied in the concept of sustainable development. The efforts to develop an understanding of sustainable development, which started with the United Nations Conference on the Human Environment (UNCHE) held in Stockholm in 1972, matured with the "Our Common Future" report published in 1987 by the United Nations World Commission on Environment and Development (Bozkurt & Bıçkı, 2016). The report in question defined sustainable development as development that meets the needs of the present without compromising the ability of future generations to meet their own needs (World Commission on Environment and Development [WCED], 1987, p. 43).

Defining sustainable development in this way has positioned the concept as a problem of value, responsibility and awareness that should be transferred between generations, beyond being a development perspective that can be explained in purely economic terms (Bozkurt & Bıçkı, 2016, p. 110). For this purpose, international organizations, especially UNESCO, are working on the development of sustainability and environmental awareness through education. In the 10-year roadmap of the UNESCO Education for Sustainable Development program in 2020, it is stated that it will not be possible to cope with the global climate crisis without bringing about fundamental changes in the way of thinking and behavior of human beings through education (UNESCO, 2020).

Türkiye is located in the Mediterranean Basin, which is among the regions that will be most affected by the negative effects of possible climate change. It is stated that the possible impacts of climate change in Türkiye, some of which are already being seen, will be in the form of increased summer temperatures, decreased winter precipitation in western provinces, loss of surface waters, increased frequency of drought, soil degradation, coastal erosion and flooding. It is predicted that climate change will cause adverse effects such as decrease in water resources, increase in floods, forest fires, drought and desertification and ecological deterioration caused by them. In the *Gediz* and *Büyük Menderes* Basins, which are among the important agricultural areas of Türkiye, estimations show that 50% of the surface water will be lost by the end of this century and water shortages will occur in agriculture, domestic field and industry (Climate Change National Action Plan [IDUEP], 2012).

Awareness of the environmental issue at the institutional level in Türkiye dates to the 1970s. In 1978, it was included in the state policy with the establishment of the Prime Ministry Undersecretariat of Environment to deal with national and international activities related to the environment (see Özmehmet, 2008). The protection of the environment was included in the 1982 Constitution for the first time. The purpose of the Environmental Law No. 2873, which entered into force in 1983, is to protect the environment, which is the common property of all living things, in line with the principles of sustainable environment and sustainable development. In the law, sustainability is expressed as the process of improving, protecting and developing all environmental values without compromising the existence and quality of resources that future generations will need. The law also contains the inclusion of environmental issues in the curriculum of the Ministry of National Education in order to raise awareness of environmental protection.[2]

Türkiye has included measures to combat climate change in its national development plans in accordance with the "common but differentiated responsibilities" principle of the United Nations Framework Convention on Climate Change. To reduce the negative effects of global climate change and to adapt to these effects, it

[2] "Enviromental Law No.2873", Resmî Gazete. 18132, 11.8.1983. https://www.mevzuat.gov.tr/anasayfa/MevzuatFihristDetayIframe?MevzuatTur=1&MevzuatNo=2872&MevzuatTertip=5. For further information see Bozkurt and Bıçkı (2016).

has been determined as the strategic aim of the Ministry of Environment and Urbanization to improve the national preparedness level by preparing strategy documents. Among the studies planned to be carried out are the joint efforts of the public, private sector, universities, non-governmental organizations, etc., and studies to change consumption patterns in a climate-friendly way (IDUEP, 2012). Therefore, the aim of raising environmental awareness through education and studies for this purpose are included in Türkiye's climate change strategy document.

7.3 Sustainability and Education

Mangunjaya and McKay (2012), evaluating the regulations made by official institutions for the protection of the environment as top-down regulations, state that this method cannot provide sufficient protection. To compensate this, it is necessary to realize the development of consciousness from the bottom up and to develop an opinion about why individuals should protect the environment. The effective way to do this is to include the aims and contents of sustainability awareness in the content of education and in particular religious and moral education in societies such as Türkiye,[3] where traditional values are still strong. Because the education system has a vital role to play in promoting better understanding of the aim of sustainable development, fostering a sense of individual and collective responsibility, and thereby encouraging changes in behavior is of paramount importance (Commission of the European Communities [COM], 2001, s. 8).[4]

According to United Nations' Report of Decades for UN's Decade of Education for Sustainable Development Education, sustainable development is fundamentally about values, with respect at the center: respect for others, including those of present and future generations, for difference and diversity, for the environment, for the resources of the planet we inhabit. Education enables us to understand ourselves and others and our links with the wider natural and social environment, and this understanding serves as a durable basis for building respect. Therefore sustainability should be embedded across the curriculum with an emphasis on society environment and economy with culture as an underlying dimension (UNESCO, 2006).

---

[3] See: World Values Survey. (2023, February 3). *The Inglehart-Welzel World cultural map*. http://www.worldvaluessurvey.org/

[4] Some studies examining the relationship between sustainability awareness and education through field data are not promising enough. In a study examining the effect of environmental education in schools in Vietnam, it was found that students' knowledge levels about environmental education increased before and after education (Phan Hoang & Kato, 2016). In a study examining the attitudes and values towards the environment in Türkiye by making use of the World Values Survey data set, it was determined that a group including higher education students and intellectual workers gave priority to environmental protection at the expense of economic growth and unemployment compared to other groups. The conclusion reached throughout the research is that education does not automatically lead to environmental behavior, but rather strengthens environmental tendencies by increasing knowledge about the environment (Bozkurt & Bıçkı, 2016).

Sustainability in the social sphere focuses on the development of a participatory and democratic system in which opinions are expressed despite differences. Environmental sustainability aims to raise awareness of the fragility of natural resources and the physical environment, and the impact of human decisions and actions on them. Sustainability in the economic field prioritizes the development of an understanding that takes into account the effects of development on the environment and social structure. It is not possible to complete the awareness of sustainability without evaluating the three developments in education together. Because cultural values, knowledge, acceptance and worldviews are decisive on the content, structure and outputs of education for sustainable development (UNESCO, 2020). Caring about animate and inanimate beings is a matter of wholeness of the personality. This situation makes religious education and training important in terms of sustainability and environmental awareness, as well as other learning areas.

7.4 Traces of Sustainability in Islamic Theology

The Qur'an often directs people's attention to nature through the ontological bond between God and being and the ethical responsibility imposed on mankind. The Qur'an informs that when a person forgets the ontological bond, he will fall into a crisis of belief, and when he neglects ethical responsibility, in this world he will begin to see the consequences of the destruction and evil he has caused: *"...corruption has appeared on land and in the sea as an outcome of what men's hands have wrought: and so Allah will let them taste the evil of some of their doings, so that they might return to the right path"* (Qur'an, 30/41). In the interpretation of the verse, Muhammed Assad states that people are warned that they will disrupt the ecological balance as a result of their careless and materialistic attitudes, which will threaten their own existence.[5]

Attempting to determine the foundations of Islamic environmental theology based on the views of contemporary Islamic scholars, Mangunjaya and McKay (2012) state that the Islamic faith constitutes three main support areas for environmental awareness education. Accordingly, (i) Islamic thought establishes a religion-based environmental ethics by positioning the environment as an important element of religious thinking and lifestyle. The creation, splendor and balance of the heavens and the earth, and the suggestion that all living creatures are ummahs in the Qur'an are clear signs of this ethic. (ii) It creates a doctrine that prevents evil against them by inculcating the idea that all beings are equal in being the act of Allah. (iii) The five fundamental theory of purpose (Maqāṣid al-Shar'īah), considered by Islamic scholars to be the foundation of Islamic ethics and law, regards the

[5] Kuran.gen.tr. (2023, January 1). *Ar-Rum Suresi*. https://kuran.gen.tr/ar-rum-suresi-english-commentary-by-muhammed esed?x=s_main&y=s_middle&kid=33&sid=30

preservation of "life" as the primary goal of religion, without discriminating between living creatures.

Qadir and Zaman (2019) states that in Islamic thought, the attribute of man as a trustee/caliph in the world, the instillation of a humble (zuhd) life, and the Maqāṣid theory form the basis of a sustainable environmental theology. In Islam, sustainability of five things (life, faith, mind, property, and family) is considered to be an objective of the law (Maqāṣid al-Shar'īah) and Muslims are duty-bound by their religion to ensure the sustenance of these five objectives. This means that the Islamic development is endogenously sustainable since preservation of life is an explicit objective of the Islamic law, which makes only those development efforts Islamically permissible that do not compromise the ability of future generations to meet their own needs.

Mangunjaya and McKay (2012) determined that the Qur'an's approach to the environment is gathered under six basic concepts as a result of the workshops he conducted with hundreds of ulama (religious scholars), ustaz (Islamic teachers), imam (religious leaders) and khatib (preachers) who research the Qur'an in different regions of Indonesia: (i) Tawhid: Because of the oneness of being, each of its member is sacred and worthy of protection. (ii) People: Every part of the being deserves attention as it is the creation of Allah. (iii) Trial Balance: The being is created with a measure and balance that human beings are expected to maintain. (iv) Ihsan: The fact that Allah has created everything with ihsan (beauty and care) and demands the same attitude from mankind. (v) Fasad: The awareness of man that he has a destructive capacity that can cause the disaster of the environment. (vi) Khalifa: Motivating to develop a sense of responsibility towards the potential for corruption (fasad) in man. The concepts briefly express the responsibilities imposed on human beings toward the environment in Islamic thought.

According to Ardoğan (2012), the understanding of existence in Islam proposes the following principles for a sustainable environment: (i) Developing world peace as the basic principle of religion (ii) Acting with the awareness that we are not the master of existence, but a part of it in creation. (iii) To protect all forms of life – biodiversity. (iv) To treat all living things with compassion. (v) Avoiding waste (vi) Being clean. The theological ground on which these principles are based is that the real owner of the property is God, and that human beings are given the role of only a "trustee". The individual cannot see the earth as a property that he can dispose of as he pleases. When a person does not act accordingly, he or she is subject to the concept of cruelty. In the broadest sense, the concept of cruelty is "to act against the creation/fitra of a being".

Islamic theology creates a human profile that Allah has appointed as a trustee to benefit from the blessings of the world. This profile builds a moral vision that can be summarized in the responsibility of maintaining balance and measure against man himself and the entire being. This morality, which we can describe as the Islamic environmental morality, sees man as the caliph, trustee and protector of nature, not the ruler, and imposes a special moral responsibility on him. In addition, in Islamic metaphysics, the universe also has a spiritual dimension. The word "world/âlem", which is used to express the whole existence, comes from the root of "ilm", which means to know, and is defined as beings other than Allah, with whom the Creator is known (Aydın, 2014, p. 551). Therefore, the meanings that can be understood by the contemplation that being means and points actually constitute the essence of religion.

Yaran (2008) structured the findings we have explained about environmental ethics in Islamic theology as a very systematic and functional model. According to him, Islamic environmental ethics has built four basic approaches and they suggest a hierarchical attitude towards the environment. The four basic approaches are: usefulness, responsibility, virtue and wisdom. All four approaches constitute altogether Islamic environmental ethics theory, and all of them are based on the teachings and advice of the Qur'an and Sunnah. However, it is the order of priority of the concepts and actions they care about that ensures differentiation. While some approach the environment on the axis of benefit-harm, some emphasize the awareness of duty and responsibility more. Others center virtue or wisdom.

These four approaches, which reveal the understanding of environmental ethics in Islam, are classified hierarchically. While usefulness expresses the lowest level in relations with the environment, responsibility expresses a protective attitude beyond it, and wisdom expresses an understanding that sees the being as valuable in its own essence (Yaran, 2008). Therefore, in the development of environmental awareness based on the Islamic belief, developing an attitude in accordance with the responsibility approach, virtue and wisdom is considered more ideal rather than usefulness approach. It is possible to assume a similar hierarchy for an environmental awareness education integrated with religious education. It can be expected that the aims and contents of the curriculum will contribute more to responsibility, virtue and wisdom instead of usefulness. In this respect, Islamic environmental ethics theory which is revealed by Yaran provides a functional framework for tracing sustainable environmental awareness education in Religious Culture and Ethics Knowledge[6] curriculum and textbooks.

Three major revisions have been carried out in Religious Culture and Ethics Knowledge curricula, which have been a part of the compulsory curriculum in Türkiye since 1982, valid between 1982–1999, 2000–2017 and 2018 to the present. The human-environment relationship took place in different forms among the aims, values, learning outcomes and subjects of these curricula. In the curricula, the subject is handled within the framework of titles such as Environmental Cleanliness, Human-Existence Relationship and Environmental Awareness in Religions, and human responsibilities to existence, protection of natural balance, measurement and environmental awareness in production-consumption relations. This research discusses how the subject is handled in the textbooks as the basic teaching tools in which the aims, learning outcomes and subjects in the curriculum are embodied and reflected on the students. In this context, the aim of the study is to evaluate the contribution of religious education to sustainable environmental education in Türkiye from 1982 to the present through Religious Culture and Ethics Knowledge curriculum and textbooks.

---

[6] Din Kültürü ve Ahlak Bilgisi (DKAB).

7.5 Method

The research is a document review structured in accordance with the qualitative research approach. In the study, primary and high school Religious Culture and Ethics Knowledge curriculum and textbooks were subjected to content analysis, which were published in 1982, 2010 and 2018, representing the fundamental curriculum revisions since the Religious Culture and Ethics Knowledge course became compulsory. In the content analysis, Yaran's model, which is considered as a functional summary of Islamic environmental theology/theory is taken as reference. In this way, it has been tried to understand whether the curriculum and textbooks have the perspective of usefulness-responsibility or virtue-wisdom in terms of the environment. In addition, clues that will enable evaluations to understand the changes in the process were tried to be found.

The themes and codes used for content analysis in the research are derived from the principles proposed by Yaran (2008) for each approach and explanations of the principles.[7] The themes, expansions and codes used in our research summarized in Table 7.1. For instance, in usefulness, nature is examined first of all on the axis of benefit-harm concepts. The main consideration in benefit and harm is the benefit and harm of human beings. This approach is largely anthropocentric, although there is no strict or extreme version of it. The usefulness has two basic principles: One of them is the "blessing principle", which states that nature has many benefits for man; the other is the "verse principle", which expresses the opportunity to draw a lesson from nature.

The responsibility is based on looking at the beings around us in terms of their rights and our duties, obligations and responsibilities towards them, rather than whether they are for our benefit or not. Regarding the responsibility approach, two basic principles draw attention or stand out: trust and caliphate. The environment is entrusted to us, and it is our duty and responsibility to protect this trust and not to betray it. In addition, our responsibility does not only end with protecting the trust, but being its successor also imposes high level duties and responsibilities (Yaran, 2008, p. 126).

Above the responsibility approach is the virtue approach. Virtue approach means taking care of the environment not as a duty or responsibility but as a voluntary attitude. Doing good deeds beyond responsibility, having love and making renunciations are matters of morality and virtue. This attitude towards the environment is called virtue. Compassion and love for all beings are the two basic principles of this approach. Wisdom, which is the highest approach, is the name of superior qualities such as being able to act as required by knowing the background of beings, events and facts, but at the same time being modest about the limit of one's knowledge. The two basic principles of this level are to be aware that the being worships the creator in its own way and has a sacredness that comes from being the work of the creator (Yaran, 2008, p. 126).

[7] See Table 7.1.

Table 7.1 Themes, explanations and codes used in content analysis

Approach	Basic claim	Sub-principles/ Codes	Supporting words/ Codes
Usefulness	Everything in the universe has been put at the service of man as a blessing and means of lesson. Anthropocentric thinking is essential.	Blessing	Do not pollute/ Keep it clean Do not waste
		Verse	Think Draw a lesson
Responsibility	Man has a responsibility before Allah and conscience towards the entire universe/ environment, which is presented to him as a blessing.	Trust	Do not betray/ harm Protect
		Khalifa	Improve Win the test
Virtue	It is essential to approach the being with a voluntary love and compassion beyond a simple responsibility.	Compassion	Do not be cruel/Do not oppress Be compassionate
		Love	Love Favor
Wisdom	To be able to look at the beings in the universe with a search for wisdom above everyone's eyes.	Servitude	See as a worshipper/servant Compete with one another in taqwa
		Sanctimony	Regard it as a masjid Regard it sacred

Adapted from Yaran (2008)

The study first examined the expressions, titles, and learning outcomes that were evaluated to provide contributions to environmental awareness and sustainability in the introduction sections and unit contents of the curriculum, in chronological order. Then, relevant textbooks for the corresponding class level were examined based on the findings from the curriculum. Instead of examining textbooks for all class levels, textbooks for the class level where relevant findings were identified in the curriculum were included in the examination. Textbooks for class levels where no findings were identified in the curriculum were not included in the examination. Considering that the official views represented by the Ministry of National Education's own publications would have a higher representational capacity, textbooks published by the Ministry were included in the examination, while the private sector was excluded.

The data collection process was carried out in two stages. In the first stage, the authors analysed the texts independently, in the second stage, the findings of the first stage were compared and final coding was made together. The main elements of the Islamic environmental theory were examined in the texts in terms of words or sentences or paragraphs, so these parts of the text were considered as analysis units. Determining the frequency of the words in the text was not satisfied, but considering the context of the relevant code, it was interpreted which approach it was associated with. If the codes given in Table 7.1 are mentioned literally in the text, this situation is given as “direct relevance” in the tables, and if this result is derived from the interpretation of the meaning integrity, it is given as “indirect relevance”. The examined textbooks and their bibliographic information are provided in Table 7.2.

7.6 Findings

7.6.1 Sustainability in 1982 Curriculum and Textbooks

In the introduction section of the 1982 Primary Education (grades 4–8) curriculum, which explains the general objectives and philosophy, statements related to usefulness and virtuousness from the four approaches of Islamic environmental theory were identified. The objectives of the curriculum included mentioning the concepts of cleanliness and external cleanliness, stating that these are among the principles of religion, and expressing that gaining this awareness among students is among the lesson objectives (Ministry of National Education [MEB], 1982a, p. 10).

This aim, which is evaluated under usefulness in the Islamic environmental theology, is included in the unit titles and sub-titles of the curriculum in the 5th, 6th, and 7th grade levels with concepts such as cleanliness, environmental cleanliness, thriftiness, and avoiding extravagance. These concepts are included in a total of four unit titles and ten sub-titles at three grade levels. Although there are no expressions about the responsibility approach in the introduction section of the curriculum, this approach is also included in two sub-titles in the fourth and sixth grades in the content. The way in which the responsibility is included in subtitles is concretized with expressions such as “protecting nature and animals.” (MEB, 1982a, pp. 6, 14).

The general aim statement “to feel respect, love, and compassion towards all beings” included in the curriculum is consistent with the virtue approach in Islamic environmental theory (MEB, 1982a, p. 10). This aim statement has been addressed in four different subheadings in the ninth unit of the fourth grade and the eighth unit of the sixth grade. Concepts such as loving nature, loving animals, loving trees and forests, and loving natural beauties are included in the relevant units’ subheadings (MEB, 1982a, pp. 6, 13, 14).

At the high school level (grades 9–11), only the seventh unit of the ninth grade covers the topic of “wastefulness.” However, the learning objectives and subtopics mostly focus on the societal benefits of avoiding wastefulness, in line with the principle of usefulness, rather than solely on the individual. This demonstrates a perspective in accordance with the principles of responsibility and virtue (MEB, 1982b, p. 23). The data shows that in the 1982 primary education curriculum, Islamic environmental theology was largely taught through the principles of usefulness, responsibility, and virtue. Table 7.3 illustrates how this perspective is reflected in the textbooks.

The data in Table 7.3 shows that in the textbooks prepared according to the 1982 Religious Culture and Ethics Knowledge (Primary Education Council) curriculum,

Table 7.2 Textbooks included in the review

Curriculum	Reviewed textbooks	Excluded
1982	Yavuz, K., & Günay, Ü. (1982). İlkokullar için Din Kültürü ve Ahlak Bilgisi 4. Sınıf. Murat Matbaacılık Koll. Şti. Bilgin, B. (1982). İlköğretim Din Kültürü ve Ahlak Bilgisi 5. Sınıf Ders Kitabı. Tifdruk Matbaacılık Sanayii A. Ş. Şener, A., & Karmış, O. (1986). Ortaokullar için Din Kültürü ve Ahlak Bilgisi 1. Tunç, C. (1982). Ortaokullar İçin Din Kültürü ve Ahlak Bilgisi 2. Sınıf Ders Kitabı. İhlas Matbaacılık ve Dağıtım A.Ş. Ayas, M. R., & Tümer, G. (1982). Lise ve Dengi Okullar için Din Kültürü ve Ahlak Bilgisi 1.Sınıf Ders Kitabı. Emel Matbaacılık Sanayii.	8th grade 10th grade 11th grade
2010	Akgül, M., Aldı, M., Çiftçi, T., Ekşi, A., Koç, E., Paça, H., Türker, A. S., Albayrak, A., Çatal, A., Demirci, F., Kara, A., Özbay, E., & Şimşekçakan, M. (2011a). İlköğretim Din Kültürü ve Ahlak Bilgisi Ders Kitabı 4. Sınıf. Millî Eğitim Bakanlığı Yayınları. Akgül, M., Albayrak, A., Çatal, A., Demirci, F., Kara, A., Özbay, E., Türker, A. S., Yıldırım, R., Aldı, M., Çiftçi, T., Ekşi, A., Koç, E., & Paça, H. (2011b). İlköğretim Din Kültürü ve Ahlak Bilgisi Ders Kitabı 5. Sınıf. Millî Eğitim Bakanlığı Yayınları. Akgül, M., Albayrak, A., Çiftçi, T., Kara, A., Özbay, E., Türker, A. S., Yıldırım, R., Çatal, A., Ekşi, A., & Şimşekçakan, M. (2011c). İlköğretim Din Kültürü ve Ahlak Bilgisi Ders Kitabı 7. Sınıf. Millî Eğitim Bakanlığı Yayınları. Akgül, M., Albayrak, A., Çiftçi, T., Kara, A., Paça, H., Türker, A. S., Yıldırım, R., Çatal, A., Ekşi, A., Koç, E., & Şimşekçakan, M. (2011d). İlköğretim Din Kültürü ve Ahlak Bilgisi Ders Kitabı 8. Sınıf. Millî Eğitim Bakanlığı Yayınları. Türkan, A., Şahan, R., Meydan, A., & Türker, A. S. (2012). Ortaöğretim Din Kültürü ve Ahlak Bilgisi Ders Kitabı 9. Sınıf. Millî Eğitim Bakanlığı Yayınları. Akgül, M., Albayrak, A., Çatal, A., Çiftçi, T., Ekşi, A., Yıldırım, R., Kara, A., Koç, E., Özbay, E., Paça, H., & Türker, A. S. (2012). Ortaöğretim Din Kültürü ve Ahlak Bilgisi Ders Kitabı 12. Sınıf. Millî Eğitim Bakanlığı Yayınları.	6th grade 10th grade 11th grade
2018	Yiğit, H., Doğan, E. Ö., Kırman, S., Özüdoğru, F., & Menküç, D. (2018). İlkokul Din Kültürü ve Ahlak Bilgisi 4 Ders Kitabı. Millî Eğitim Bakanlığı Yayınları. Yiğit, H., Açık, A., Doğan, E. Ö., Dirlik, E. E., & Macit, A. (2019). Ortaokul Din Kültürü ve Ahlak Bilgisi 5. Millî Eğitim Bakanlığı Yayınları. Nayir, S., Açık, A., Beyazal, M., & Paksoy, H. B. (2022). Ortaokul Din Kültürü ve Ahlak Bilgisi 8 Ders Kitabı. Millî Eğitim Bakanlığı Yayınları. Konaklı, N., Çınar, H., & Emiroğlu, S. (2021). Ortaöğretim Din Kültürü ve Ahlak Bilgisi 10. Millî Eğitim Bakanlığı Yayınları. Yılmaz, F., Doğan, H. İ., Özkan, S., & Öztürk, Y. (2022). Ortaöğretim Din Kültürü ve Ahlak Bilgisi 12. Millî Eğitim Bakanlığı Yayınları.	6th grade 7th grade 9th grade 11th grade

Table 7.3 Sustainable environmental consciousness in books written according to the 1982 curriculum

Approach	Principles	Expression	Directly	Indirectly	Total	Total	Total
Usefulness	Blessing	Do not pollute	10	2	12	21	24
		Do not waste	2	7	9		
	Verse	Think	1	1	2	3	
		Draw a lesson	1	–	1		
Responsibility	Trust	Do not harm	4	1	5	13	16
		Protect	8	–	8		
	Caliphate	Improve	3	–	3	3	
		Win the test	–	–	0		
Virtue	Compassion	Do not be cruel	–	–	0	0	4
		Be compassionate	–	–	0		
	Love	Love	3	–	3	4	
		Favor	1	–	1		
Wisdom	Servitude	See as a servant	–	–	0	1	3
		Compete with one another in taqwa	–	1	1		
	Sanctimony	Regard it as masjid	1	–	1	2	
		Regard it sacred	1	–	1		

the approaches that make up Islamic environmental theory are listed in order from most to least prevalent as usefulness, responsibility, virtue, and wisdom. Expressions and recommendations such as "not polluting the environment, air and water", "not wasting blessings", "contemplating creation", and "learning from living and non-living things" in the textbooks aim to direct the student's attention to nature and develop an awareness of it, either directly or indirectly. However, since the usefulness approach teaches the human to see the environment as a blessing and a sign, it can be evaluated as indirectly attributing value to the environment. The usefulness is considered the weakest contribution to sustainability among the four approaches hierarchically. In contrast to the usefulness, the responsibility approaches, which emphasizes the responsibility of protecting nature and making it better, has a smaller place in the books. The virtue and wisdom, which see nature as inherently valuable and expect to be treated with love and sanctity, have very little place in the books.

7.6.2 Sustainability in 2010 Curriculum and Textbooks

The Religious Education and Ethics Knowledge course, which was made mandatory in 1982, had a subject and teacher-centered approach reflecting traditional teaching methods. With revisions made in 2000, 2005, 2006, and 2010, efforts were made to transition to a constructivist, student-centered, and diversity-oriented approach. An examination of the 2010 primary school curriculum showed that expressions aimed at developing environmental awareness were included in the vision and goals section to represent this new approach. The curriculum's vision includes statements about raising individuals who are "sensitive to the environment" and "protect cultural heritage and nature." The curriculum's goals also include "developing awareness of physical and social environment protection" (MEB, 2010a, pp. 8, 13). Units aimed at developing environmental awareness were included in the fourth, seventh, and eighth grades. Expressions such as "keeping the environment clean, protecting the environment, loving nature, loving animals, and protecting nature" were used in the learning outcomes and subheadings.

The high school curriculum's vision in 2010 repeats the expression "sensitive to the environment" found in the primary school curriculum. The curriculum's goal section includes the expression "reaches awareness of the environment, natural balance, and cleanliness" (MEB, 2010b, pp. 9, 12). When examining the content of the units, learning outcomes and headings aimed at developing environmental awareness were found in one unit each in the ninth and twelfth grades. Learning outcomes and headings were created using concepts such as "environmental cleanliness, the importance of cleanliness in religion, the importance of protecting the environment in religions, and order in the universe." The findings show that the responsibility approach stands out in the primary school curriculum, while the usefulness approach stands out in the high school curriculum compared to the 1982 curriculum. There was no finding related to the virtue and wisdom. Additionally, it can be said that clearer expressions related to environmental awareness were included in the general goals of the curriculum in 2010 compared to the 1982 curriculum. The reflection of the curriculum's approach in textbooks is given in Table 7.4.

The most frequently discussed concept among the approaches that comprise Islamic environmental theory in textbooks prepared according to the 2010 Religious Culture and Ethics Knowledge Curriculum is responsibility. This aligns with the emphasis on responsibility in the primary education curriculum, where the approach is covered with statements such as "protecting nature and animals is a religious duty", "protecting the right to life and reproduction of living beings", "all created beings are entrusted to humans", and "preserving air and water resources". Faith is presented as the source of responsibility in the coverage of these topics. The emphasis on the usefulness, which highlights the benefits derived from nature and living beings, can be evaluated as a positive development for the prominence of responsibility in protecting nature and animals. In addition, the fact that the virtue and wisdom approaches, which acknowledge that there is inherent value in existence regardless of its benefit to humans, are emphasized in a way that is close to usefulness and responsibility compared to the 1982 curriculum, is also a positive development. Qualitatively, a more pronounced view that represents a higher point of Islamic environmental theology is expressed in the 2010 curriculum. However, it is also noteworthy that textbooks prepared according to the 2010 curriculum emphasize environmental consciousness less frequently than those prepared according to the 1982 curriculum.

Table 7.4 Sustainable environmental consciousness in books written according to the 2010 curriculum

Approach	Principles	Expression	Directly	Indirectly	Total	Total	Total
Usefulness	Blessing	Do not pollute	1	–	1	1	4
		Do not waste	–	–	0		
	Verse	Think	–	1	1	3	
		Draw a lesson	1	1	2		
Responsibility	Trust	Do not harm	3	2	5	8	12
		Protect	3	–	3		
	Caliphate	Improve	3	–	3	4	
		Win the test	1	–	1		
Virtue	Compassion	Do not be cruel	2	–	2	3	3
		Be compassionate	1	–	1		
	Love	Love	–	–	0	0	
		Favor	–	–	0		
Wisdom	Servitude	See as a servant	1	–	1	1	7
		Compete with one another in taqwa	–	–	0		
	Sanctimony	Regard it as masjid	–	–	0	6	
		Regard it sacred	5	1	6		

7.6.3 Sustainability in 2018 Curriculum and Textbooks

The Ministry of National Education made significant changes to the teaching curricula of many subjects, including the Religious Culture and Ethnic Knowledge course, in 2018. The most notable feature of the changes is the simplification of curriculum content and aims, criteria, reduction of unit and learning outcomes numbers, and inclusion of current religious issues in the curriculum. Additionally, while maintaining a constructivist and student-centered approach, skill-based education has also been utilized. It is understood that the simplification of curriculum texts has resulted in a negative outcome in terms of expressions related to environmental consciousness and sustainability. In our research, we found no expressions related to environmental consciousness and sustainability in the introduction (vision/philosophy, goals, and principles) of primary (grades 4–8) and high school (grades 9–12) curricula.

When the curricula are analysed in terms of unit headings, subheadings, and learning outcomes, it is found that in the primary, there is content related to environmental consciousness in one unit for fourth and fifth grades and two units for eighth grade (Ministry of National Education, 2018a). The found content is patterned around the concepts of "environmental cleanliness", "the meaning of the perfect order in the universe", "the measure and destiny of existence", and "compassion and love." The concepts mostly demonstrate the usefulness of the topic (in the context of verses and blessings). In the high school curriculum, the subject is included in a total of four units, with unit headings, subheadings, and learning outcomes included in the tenth and twelfth grades. There is a unit titled “Religion and Environment” in the tenth grade (Ministry of National Education, 2018b). The unit learning outcomes indicate that it is aimed for students to understand the contributions of religion to the preservation of the environment. In the twelfth grade unit, “Current Religious Issues”, learning outcomes related to current environmental and sustainability issues are also included. Here, students are expected to evaluate food safety and genetically modified food topics from a religious perspective. The reflection of the curriculum’s approach in textbooks is given in Table 7.5.

When examining the reflection of approaches in Islamic environmental theory in textbooks, it is seen that the usefulness approach, which evaluates the environment in terms of its benefits to humans, is strongly emphasized. While the responsibility approach, which expresses a much more ethical understanding in dealing with the environment, is less discussed, the virtue approach is discussed only in one phrase. However, there is no place for a wise Islamic perspective on the environment. This situation shows that the sterility in the aims, philosophy, and principles of the curricula is reflected in the textbooks, even though the current curriculums are the ones that give the clearest place to environmental issues in their unit content. In the textbooks, the understanding that positions Islam in relation to the environment as a tool that serves to ground or sustain human belief dominates. The approaches that

Table 7.5 Sustainable environmental consciousness in books written according to the 2018 curriculum

Approach	Principles	Expression	Directly	Indirectly	Total	Total	Total
Usefulness	Blessing	Do not pollute	1	3	4	8	9
		Do not waste	4	–	4		
	Verse	Think	–	1	1	1	
		Draw a lesson	–	–	0		
Responsibility	Trust	Do not harm	–	–	0	2	3
		Protect	1	1	2		
	Caliphate	Improve	1	–	1	1	
		Win the test	–	–	0		
Virtue	Compassion	Do not be cruel	–	–	0	1	1
		Be compassionate	1	–	1		
	Love	Love	–	–	0	0	
		Favor	–	–	0		
Wisdom	Servitude	See as a servant	–	–	0	0	0
		Compete with one another in taqwa	–	–	0		
	Sanctimony	Regard it as masjid	–	–	0	0	
		Regard it sacred	–	–	0		

prioritize ethical duties towards all existence, such as responsibility, and attribute value to existence in its essence, such as virtue and wisdom, have found very little place.

Despite the existence of an independent unit titled "Religion and Environment" in the 2018 Curriculum and the requirement to examine environmental issues from a religious perspective under the unit of current religious issues, it has been observed that the content related to environmental awareness is quite limited in the textbooks prepared according to these curricula. This situation can be considered as a result of the fact that the aims of the curricula have not been sufficiently reflected in the textbooks, or in general, it can be evaluated in relation to the tendency to approach the topic from a theological perspective based on classical teachings, even for current issues in DKAB classes. On the other hand, it has been determined that the frequency of content related to environmental awareness is lower compared to textbooks prepared for previous curricula. This situation can be evaluated in connection with the simplification of the curricula, which resulted in reducing the number of units and titles.

7.7 Conclusion and Evaluation

Türkiye is one of the countries expected to be most affected by the negative impacts of climate change. It is also a country with a majority Muslim population, where religious values hold an important place in society. This situation makes the path to be followed in the teaching of religious values critical for the development of sustainable environmental consciousness. From the perspective of Islamic theology, the hierarchy of usefulness, responsibility, virtue, and wisdom approaches points to a classification from weakest to strongest in dealing with the environment. Usefulness indicates the understanding that existence was created for the benefit of humans, responsibility refers to passive protection of other beings by humans, virtue represents voluntary kindness towards existence, and wisdom represents a perspective that considers existence as valuable in its essence. Therefore, the reflection of this environmental theory found in Islamic sources in the teaching curricula and textbooks gives us an idea about religious education in sustainable environmental consciousness.

In the examination we conducted in the introductory sections of the Religious Culture and Ethics Knowledge course curricula in three different periods from 1982 when the course was compulsory to the present day (vision, philosophy, objectives, and principles), we encountered statements about the contribution of the course to sustainable environmental consciousness in the 1982 and 2010 curricula. The approach to the subject in these curricula evokes the perspective of usefulness, which is the weakest link of Islamic environmental theology. However, no explanation about environmental consciousness was found in the introduction of the 2018 Religious Culture and Ethics Knowledge curriculum currently in force. Although it is possible to explain this situation with the simplification made in the curriculum texts, sacrificing such a vital issue to simplification is not considered a healthy attitude. Considering that the introductions of educational curricula are documents that reveal the official educational philosophy, it can be understood that the Ministry of National Education (MEB) could not establish a healthy relationship between environmental consciousness and lessons. Indeed, Tanrıverdi (2009), who examined the place of sustainable environmental understanding in the general primary school curriculum in Türkiye, determined that the concept of the environment was limited to the science and technology area and lower-level acquisitions.

The Climate Change Strategy Document for Türkiye sets the goal of organizing educational, awareness-raising, informative, and public awareness-raising activities to develop the capacity to combat and adapt to climate change (IDUEP, 2012, p. 58). However, although the same document mentions collaboration with organizations from various fields such as NGOs, industry, and agriculture in carrying out sustainability efforts, it does not provide a clear statement regarding collaborations that can be made with the Ministry of National Education. The subject matter of the strategy document is mainly constructed around educational activities aimed at teaching new techniques for sustainability in agriculture and industrial production. However, it is understood that there is also a consumption aspect to the issue, and it is not given enough consideration that this aspect is related to the development of care for nature in all individuals, not just those involved in the production process.

Examinations of the unit learning outcomes and contents included in the curricula have shown that environmental consciousness and environmental ethics themed topics have been included in all three periods. The approach to the topics is mostly based on concepts such as cleanliness, waste reduction, understanding the order in the universe, drawing lessons, benefiting from blessings, sensitivity to the environment in which one lives, and showing love and compassion to living beings. It is not unnoticed that the beneficial perspective is dominant in the approach to the environment in all curricula. The 2018 curriculum stands out from previous curricula with its clear and up-to-date unit learning outcomes and titles regarding sustainable environmental consciousness. In the 2018 curriculum, religion and the environment are addressed as a separate topic, and environmental and food issues are included as a subheading among current religious issues.

When it comes to the reflection of curriculum content in textbooks, it is not possible to make the same optimistic evaluations. As content progresses from past to present in textbooks, it becomes sparser. The treatment of the subject matter in textbooks, like in curricula, is heavily focused on usefulness, and partially on the responsibility. This means that the environment is valued and protected not for its own sake, but for its religious and worldly benefits to humans. The virtue and wisdom approaches of the Islamic environmental theory, which represent higher levels, are rarely found in the books. Especially in textbooks prepared according to the 2018 curriculum, this deficiency is noticeable. In books prepared based on the 2018 curriculum, which aim to include current issues in the curriculum, the subject is approached from a theological perspective, and the environment is addressed based on the meaning attributed to it by theology, rather than the current problems of the environment and their cultural and economic causes and consequences.

Parker's (2017) analysis of current curricula in Indonesia with regards to sustain able environmental awareness and religion has revealed that the new curriculum neglects the interrelationships of economic development and environmental sustainability and frames the environment within a creationist, religious worldview. Critical analysis of the curriculum shows us that the environment most often appears as the creation of God, for which students are to be grateful. Human agency for the development of natural resources is present in the curriculum, but responsibility for the destruction of the natural world is neglected. Our research findings being similar to those in Indonesia indicate the general need for meaningful learning-based updates in Islamic education. Meaningful learning is the presentation of accumulated traditional knowledge reconsidered in light of the current reality, conveyed to the student through associations that are placed in their conceptual world (Ausubel, 2000).

Keskiner's (2014) evaluation of religious courses taught in Imam-Hatip schools, which provide vocational religious education alongside formal general education, with regards to environmental awareness, shows similar findings. It is determined that the universe is generally approached from a theological perspective, and the relationship between faith and the environment largely centers on the responsibility of taking lesson and advice for it, rather than being careful and compassionate towards it. In textbooks, the relationship between Allah and humans is emphasized over the universe, rather than the relationship between humans and the environment, and human responsibility towards the environment falls behind the idea of the environment being at the service of humans. Parallel to Keskiner's (2014) findings, our study shows that healthy relationships between current environmental issues and religious references cannot be established in religious education textbooks. We believe that this problem is related not only to the need for current interpretations of Islamic sources but also to the understanding of textbook preparation as a process of reproducing knowledge, rather than transferring it. If the presentation of knowledge in the educational process allows for reinterpretation, traditional knowledge can only provide solutions to current problems. Indeed, UNESCO states that sustainable education is not just a content-based process, but also requires a participatory transformation in the educational process (2020).

References

Akgül, M., Aldı, M., Çiftçi, T., Ekşi, A., Koç, E., Paça, H., Türker, A. S., Albayrak, A., Çatal, A., Demirci, F., Kara, A., Özbay, E., & Şimşekçakan, M. (2011a). İlköğretim Din Kültürü ve Ahlak Bilgisi Ders Kitabı 4. Sınıf. Millî Eğitim Bakanlığı Yayınları.

Akgül, M., Albayrak, A., Çatal, A., Demirci, F., Kara, A., Özbay, E., Türker, A. S., Yıldırım, R., Aldı, M., Çiftçi, T., Ekşi, A., Koç, E., & Paça, H. (2011b). İlköğretim Din Kültürü ve Ahlak Bilgisi Ders Kitabı 5. Sınıf. Millî Eğitim Bakanlığı Yayınları.

Akgül, M., Albayrak, A., Çiftçi, T., Kara, A., Özbay, E., Türker, A. S., Yıldırım, R., Çatal, A., Ekşi, A., & Şimşekçakan, M. (2011c). İlköğretim Din Kültürü ve Ahlak Bilgisi Ders Kitabı 7. Sınıf. Millî Eğitim Bakanlığı Yayınları.

Akgül, M., Albayrak, A., Çiftçi, T., Kara, A., Paça, H., Türker, A. S., Yıldırım, R., Çatal, A., Ekşi, A., Koç, E., & Şimşekçakan, M. (2011d). İlköğretim Din Kültürü ve Ahlak Bilgisi Ders Kitabı 8. Sınıf. Millî Eğitim Bakanlığı Yayınları.

Akgül, M., Albayrak, A., Çatal, A., Çiftçi, T., Ekşi, A., Yıldırım, R., Kara, A., Koç, E., Özbay, E., Paça, H., & Türker, A. S. (2012). Ortaöğretim Din Kültürü ve Ahlak Bilgisi Ders Kitabı 12. Sınıf. Millî Eğitim Bakanlığı Yayınları.

Ardoğan, R. (2012). İslam'da Çevre Teolojisinin Pratiğe Yansıması: Çevre Ahlakı. Birey ve Toplum Sosyal Bilimler Dergisi. https://dergipark.org.tr/en/pub/birtop/issue/3528/47968

Ausubel, D. P. (2000). *The acquisition and retention of knowledge: A cognitive view*. Springer.

Ayas, M. R., & Tümer, G. (1982). Lise ve Dengi Okullar için Din Kültürü ve Ahlak Bilgisi 1.Sınıf Ders Kitabı. Emel Matbaacılık Sanayii.

Aydın, M. Ş. (2014). Varlık Bilincine Dayalı Çevre Ahlakı. Ulusal Çevre ve Ahlak Sempozyumu Bildiri Metinleri, 549–557. https://www.google.com/search?q=VARLIK+B%C4%B0L%C4%B0NC%C4%B0NE+DAYALI+%C3%87EVRE&oq=VARLIK+B%C4%B0L%C4%B0NC%C4%B0NE+DAYALI+%C3%87EVRE&aqs=chrome..69i57j69i60l3.12564j0j4&sourceid=chrome&ie=UTF-8

Bilgin, B. (1982). İlköğretim Din Kültürü ve Ahlak Bilgisi 5. Sınıf Ders Kitabı. Tifdruk Matbaacılık Sanayii A. Ş.

Bozkurt, V., & Bıçkı, D. (2016). Sürdürülebilir Kalkınma, Çevre Bilinci ve Değerler. Dokuz Eylül Üniversitesi İktisadi ve İdari Bilimler Fakültesi İktisat Bölümü İktisat Çalıştayı.

Commission of the European Communities (COM). (2001). *A sustainable Europe for a better world: A European Union strategy for sustainable development*.

İklim Değişikliği Ulusal Eylem Planı (İDUEP) 2011–2023. (2012). Çevre ve Şehircilik Bakanlığı. https://iklim.gov.tr/

Keskiner, E. (2014). Çevre ve Din Eğitimi: İmam Hatip Lisesi Meslek Dersleri Kitapları Üzerine Bir İnceleme. *Journal of Values Education, 12*(27), 187–215. https://dergipark.org.tr/en/pub/ded/issue/29172/312397

Konaklı, N., Çınar, H., & Emiroğlu, S. (2021). Ortaöğretim Din Kültürü ve Ahlak Bilgisi 10. Millî Eğitim Bakanlığı Yayınları.

Kur'an-ı Kerim. (n.d.). https://kuran.gen.tr/ar-rum-suresi-english-commentary-by-muhammedesed?x=s_main&y=s_middle&kid=33&sid=30

Mangunjaya, F. M., & McKay, J. E. (2012). Reviving an Islamic approach for environmental conservation in Indonesia. *Worldviews: Global Religions, Culture, and Ecology, 16*(3), 286–305. https://doi.org/10.1163/15685357-01603006

Millî Eğitim Bakanlığı (MEB). (1982a). İlköğretim Din ve Ahlak Bilgisi Dersi (4, 5, 6, 7 ve 8. Sınıflar) Öğretim Programı ve Kılavuzu.

Millî Eğitim Bakanlığı (MEB). (1982b). Ortaöğretim Din Kültürü ve Ahlak Bilgisi Dersi (9, 10 ve 11. Sınıflar) Öğretim Programı ve Kılavuzu.

Millî Eğitim Bakanlığı (MEB). (2010a). İlköğretim Din Kültürü ve Ahlak Bilgisi Dersi (9, 10, 11 ve 12. Sınıflar) Öğretim Programı ve Kılavuzu.

Millî Eğitim Bakanlığı (MEB). (2010b). Ortaöğretim Din Kültürü ve Ahlak Bilgisi Dersi (9, 10, 11 ve 12. Sınıflar) Öğretim Programı.

Millî Eğitim Bakanlığı (MEB). (2018a). Din Kültürü ve Ahlak Bilgisi Dersi Öğretim Programı (İlkokul 4 ve Ortaokul 5, 6, 7 ve 8. Sınıflar).

Millî Eğitim Bakanlığı (MEB). (2018b). Din Kültürü ve Ahlak Bilgisi Dersi Öğretim Programı (Ortaöğretim 9, 10, 11 ve 12. Sınıflar). Millî Eğitim Bakanlığı (MEB).

Nayir, S., Açık, A., Beyazal, M., & Paksoy, H. B. (2022). Ortaokul Din Kültürü ve Ahlak Bilgisi 8 Ders Kitabı. Millî Eğitim Bakanlığı Yayınları.

Özmehmet, E. (2008). Dünyada ve Türkiye'de Sürdürülebilir Kalkınma Yaklaşımları. *Journal of Yaşar University, 3*(12), 1853–1876. https://doi.org/10.19168/JYU.48930

Parker, L. (2017). Religious environmental education? The new school curriculum in Indonesia. *Environmental Education Research, 23*(9), 1249–1272. https://doi.org/10.1080/13504622.2016.1150425

Phan Hoang, T. T., & Kato, T. (2016). Measuring the effect of environmental education for sustainable development at elementary schools: A case study in Da Nang City, Vietnam. *Sustainable Environment Research, 26*(6), 274–286. https://doi.org/10.1016/J.SERJ.2016.08.005

Qadir, J., & Zaman, A. (2019). Sustainable development viewed from the lens of Islam. *International Journal of Pluralism and Economics Education, 10*(1), 46. https://doi.org/10.1504/IJPEE.2019.10019578

Şener, A., & Karmış, O. (1986). Ortaokullar için Din Kültürü ve Ahlak Bilgisi 1.

Tanrıverdi, B. (2009). Sürdürülebilir Çevre Eğitimi Açısından İlköğretim Programlarının Değerlendirilmesi. Eğitim ve Bilim Dergisi, 34, 151.

The Intergovernmental Panel on Climate Change (IPCC). (2018). *Global warming of 1.5 °C*. https://www.ipcc.ch/site/assets/uploads/sites/2/2022/06/SR15_Full_Report_HR.pdf
Tunç, C. (1982). Ortaokullar İçin Din Kültürü ve Ahlak Bilgisi 2. Sınıf Ders Kitabı. İhlas Matbaacılık ve Dağıtım A.Ş.
Türkan, A., Şahan, R., Meydan, A., & Türker, A. S. (2012). Ortaöğretim Din Kültürü ve Ahlak Bilgisi Ders Kitabı 9. Sınıf. Millî Eğitim Bakanlığı Yayınları.
United Nations Educational, Scientific and Cultural Organization (UNESCO). (2006). *Framework for the UNDESD international implementation scheme*.
United Nations Educational, Scientific and Cultural Organization (UNESCO). (2020). *Education for sustainable development: A roadmap*. https://unesdoc.unesco.org/ark:/48223/pf0000374802.locale=en
World Commission on Environment and Development (WCED). (1987). Report of the World Commission on Environment and Development: Our common future (The Brundtland Report). *Medicine, Conflict and Survival, 4*. https://doi.org/10.1080/07488008808408783
World Values Survey. (n.d.). *The Inglehart-Welzel World cultural map*. Retrieved 3 February 2023, from http://www.worldvaluessurvey.org/
Yaran, C. S. (2008). İslam Çevre Etiğinin Dört Kuramı ve Sekiz İlkesi (F. Kayadibi, Ed.; pp. 121–133). Yalın Yayıncılık.
Yavuz, K., & Günay, Ü. (1982). İlkokullar için Din Kültürü ve Ahlak Bilgisi 4. Sınıf. Murat Matbaacılık Koll. Şti,.
Yiğit, H., Doğan, E. Ö., Kırman, S., Özüdoğru, F., & Menküç, D. (2018). İlkokul Din Kültürü ve Ahlak Bilgisi 4 Ders Kitabı. Millî Eğitim Bakanlığı Yayınları.
Yiğit, H., Açık, A., Doğan, E. Ö., Dirlik, E. E., & Macit, A. (2019). Ortaokul Din Kültürü ve Ahlak Bilgisi 5. Millî Eğitim Bakanlığı Yayınları.
Yılmaz, F., Doğan, H. İ., Özkan, S., & Öztürk, Y. (2022). Ortaöğretim Din Kültürü ve Ahlak Bilgisi 12. Millî Eğitim Bakanlığı Yayınları.

Hasan Meydan is a Professor at Sakarya University, Faculty of Theology, Department of Philosophy and Religious Sciences. He studies values education, Islamic morality and education, meaningful learning in religious education, the contribution of religious education to making sense of life, and religious education in postmodern thought and community.

Recep Kaymakcan is a Professor of religious education and currently an advisor to the Ministry of Youth and Sports in Turkey. His academic research areas: Comparative religious education, policies of religious education, values education, youth and religion, human rights and religious education, teacher training for RE.

Feyza Karaahmetoğlu is a Ph.D. student at Sakarya University, Social Sciences Institute. She is studying religion and values education in the consumer society, consumption habits of religious individuals, and the vanity culture.

Chapter 8
Reframing the Human–Nature Relationship in Worldview Education: The Case of Finnish Lutheran and Orthodox Textbooks

Harriet Zilliacus, Lili-Ann Wolff, Pia Mikander, and Arto Kallioniemi

8.1 The Need to Reframe the Human–Nature Relationship in Worldview Education

The climate and sustainability crises are existential threats that require urgent changes in all areas of human life. In many parts of the world, the human–nature relationship is problematic, and a reframing of this relationship is central to humans' ability to face and solve the crises (e.g., Riechers et al., 2021). To support the development of more sustainable worldviews and lifestyles, policy makers and researchers have assigned a key role to education. According to international educational policies, the notion of sustainability needs to be integrated into all school subjects (Zilliacus & Wolff, 2021; UNESCO, 2021). The Finnish National Core Curriculum for Basic Education 2014/2016 links the aim of sustainability to the development of an ethical stance not only to other people, but also to non-human animals and other parts of nature. This aim is expressed as fundamental to becoming a humane and educated person and reflects a change from the previous curriculum, which was more human-centred (Finnish National Board of Education [FNBE], 2016; Zilliacus & Wolff, 2021). Against this background, it is of interest to investigate how sustainability and ethical aims regarding the human–nature relationship are implemented in Finnish worldview education. The results will also be relevant to worldview education research internationally.

Worldview education is an umbrella concept for various forms of religious education and comprises education in various beliefs, cultures and ethics (Salmenkivi et al., 2022). The sustainability and climate crises raise new and emerging ethical

H. Zilliacus (✉) · L.-A. Wolff · P. Mikander · A. Kallioniemi
The University of Helsinki, Helsinki, Finland
e-mail: harriet.zilliacus@helsinki.fi

S. Windsor, O. Franck (eds.), *Intersections of Religion, Education, and a Sustainable World*, https://doi.org/10.1007/978-3-031-81809-7_8

issues relating to the human–nature relationship, as well as to global and intergenerational concerns, which are highly relevant for worldview education. Worldviews, attitudes and beliefs are intricately associated with environmental concerns and behaviour in ways that are not yet fully understood across traditions and contexts (Jenkins et al., 2018; Rissanen et al., 2023; Stevenson et al., 2013). From an educational perspective, research and policy call for worldview education that supports religious literacy and advocacy and inspires students to envision the world anew. Understanding the changeable nature of various worldviews and cultures can enable transformative change in thoughts and lifestyles. Researchers also emphasize the socio-emotional support of children (Zilliacus & Wolff, 2021; Rissanen et al., 2023).

Although worldview education research has paid increasing attention to sustainability perspectives, more research is still needed (Rissanen et al., 2023; Tomlinson, 2019; Kvamme, 2017; Martin, 2015). Through discourse analysis (Laclau & Mouffe, 1985/2001), this study focuses on how the human–nature relationship is introduced in Finnish worldview education in the initial comprehensive school years. The research question is: In what ways is the human–nature relationship discursively portrayed in Finnish Lutheran and Orthodox religious education textbooks in the 1st and 2nd grades?

Finland has a pluralistic framework, offering separate worldview education according to parental religious belonging. This study is limited to Lutheran and Orthodox education as they represent the two folk churches and the forms of education with the longest traditions in Finnish schools. The need for research on the links between worldview education and sustainability includes various religious, non-religious and Indigenous worldviews. However, the human-centredness of Christian worldviews has often been blamed for bringing about and continuing sustainability crises (including the climate crisis) (Callicott, 2013; Mayer, 2019), which also motivates focusing on Christian education textbooks.

8.2 Human Relationships with Nature in a Christian Historical Perspective

Human relationships with nature are complex and highly diverse. However, dualistic and hierarchical relations, which are present in Western and Christian worldviews and education, have come under scrutiny, particularly from a sustainability perspective. Skilbeck, who has studied human relationships with nature from a historical Judaeo-Christian viewpoint, states that humans need a revolution to understand that "nature does not revolve around man" (2021, p. 129). De Groot and van den Born's (2007) study on theological anthropologies explored visions of nature among five population groups in Canada, including Christians, Muslims, First Nations, Buddhists and secularists. The study resulted in four metaphors expressing the relationship between humans and non-human nature. They depict the human role as *Master*, *Steward*, *Partner* and *Participant*. The first two mentioned relationships largely express human-centred worldviews, whereas the latter two express more interconnected and non-hierarchical worldviews. In the Judo-Christian context, the notion of a *Master* relationship comes from a narrative of humans having dominion over the rest of creation. This reflects a sense of detachment, which is echoed in the modern scientific revolution and more generally in Western culture. Environmental philosophy usually blames Ancient Greek philosophy as well as Christianity for the negative impact that science and technology have had on the environment (e.g., Attfield, 1994). The *Old Testament* (OT) elevation of man over both the living and non-living parts of nature has strongly influenced Western culture and still does worldwide (Skilbeck, 2021). According to the OT, humans are the crowning glory of God's creation, and are thus granted the moral right to rule over (see, e.g., Attfield, 1994).

> And God said, Let us make man in our image, after our likeness: and let them have dominion over the fish of the sea, and over the fowl of the air, and over the cattle, and over all the earth, and over every creeping thing that crept upon the earth.[1]

Men (more than women) and their needs are central in this cosmic order, in which the rest of nature serves human interests. This is also obvious in the Book of Wisdom in the Apocrypha, which emphasizes humans' role in gathering knowledge about everything on earth and labelling and utilizing it (Wolff, 2011). On their journey towards heaven, the earth is a temporary lodging for this super species. However, there is also a tendency to feel gratitude towards all that God has given humans during this journey, as well as emphasizing a love of nature and a sharing of the Earth as in the Garden of Eden (Skilbeck, 2021).

Western history has seen defenders of as well as antagonists to the notion of humans as superior to other creatures (e.g., Attfield, 1994). Therefore, concern for non-human nature is also based on Christianity's pastoral matrix and humans' role as vice-shepherds caring for the rest of creation (Wolff, 2011). This role as vice-shepherds coincides with de Groot and van den Born's (2007) second metaphor of *Stewards*. A hierarchical understanding is tempered in this metaphor through a sense of care and concern but includes leadership and responsibility to look after God's gifts, which places humans above other parts of nature.

8.3 The Human–Nature Relationship in Education

Historically, the dominant human relation to non-human nature as *Master* and *Steward* has impacted education and is interlinked with the humanist tradition within education. Humanism has placed humans at the apex, neglecting the role of nature and human relationships with the rest of nature (von Wright, 1994). Education based on humanism has emphasized that the animal part of children must be educated for them to become cultural beings (Wolff, 2011). According to Zimmermann (2015), the ideals for becoming human derive from the biblical notion of the divine,

[1] Genesis 1:26, King James Bible.

in which the Incarnation is the final stage. This teleological narrative is also linked to the relationship between the human will and divine action. Granted that humans have free will, they can critically choose their own path, and decide if they want to act morally or not. However, since free will would conflict with the divine will, humans are often regarded as subordinate to the will of God, according to several religions (Goetz, 2011).

Today, many authors conducting sustainability and sustainability educational research see need for transformed worldviews supporting interconnected, non-hierarchical, and collective human–nature relationships that also include intergenerational dimensions (Mayer, 2019; Callicott, 2013; Berry, 1988; Wolff, 2022). A sustainable worldview includes human relationships with others of their own species, with other species and with non-living nature. Undeniably, a human-centred, hierarchical worldview tends to bring about exploitation, damage, and destruction of nature and livelihoods (Riechers et al., 2021).

According to de Groot and van den Born's (2007) metaphors, unlike humans as *Masters* and *Stewards*, the relationship of *Partner* expresses humans standing side by side in partnership with non-human nature. In this view, non-human nature has its own status and independent value and humans and non-human nature are in a dynamic process of interaction that can reflect a sense of equal power between them. However, this worldview raises questions whether humans can envision non-human nature as a partner or if a certain anthropomorphism is inevitable. The concept of partnership also creates tensions from a Christian perspective, as an interconnectedness and communion in the human–nature relationship abolishes the distinctiveness and superior role given specifically to humans by religion (Martin, 2015). De Groot and van den Born (2007) also suggest a fourth metaphor, that of *Participant*. From this perspective, the separation between humans and non-human nature collapses, as humans are seen as an integral part of nature, not only biologically but also spiritually. In Groot and van den Born's study, this metaphor arose from Indigenous and Buddhist perspectives, reflecting a way of thinking about the human–nature relationship through a spiritual connection, a metaphor that also challenges traditional Christian theology as well as non-spiritual worldviews.

Within education, diverse attempts have been made to overcome dualistic and hierarchical thoughts about human relationships with non-human nature so that these thoughts are more in line with the metaphors of *Partne*r and *Participant*. Among these attempts is the *biophilia* theory, which seeks to develop "the innately emotional affiliation of human beings to other living organisms" (Wilson, 1993, p. 31). In addition, there are many branches of *posthumanism*, which all belong to the philosophical turn towards new ontologies as a reaction to human-centred worldviews. Haraway (2018) uses the expression 'making kin', which implies a non-hierarchical recognition and caring about other-than-humans. New materialism, as a branch of post-humanism, invites educational approaches that encourage people to become ecologically aware and to connect things differently. This concerns humans' being-in-the-world, and their relationship with what in post-humanism is called 'more-than-humans' (Paulsen et al., 2022). Before we turn to our analysis of human–nature relationship discourses in Finnish worldview textbooks, we will clarify the basic structure of Finnish worldview education.

8.4 Finnish Worldview Education

In Finnish schools, religious education is a compulsory subject and is non-confessional in nature. In contrast to other Nordic countries, which have a common worldview subject for all students, Finnish religious education classes are organized according to religious belonging if there are at least three children whose parents belong to a registered religious community. Currently, there are ten different religious education syllabi in Finland: five for Christian denominations (Lutheran, Orthodox, Catholic, Adventist, and Pentecostal) and five for other religions (Islam, Judaism, Buddhism, Bahá'í and Krishna) (ISCKON). Students who do not belong to a religious community, attend Ethics (in Finnish: *elämänkatsomustieto*) (Kavonius, 2021). Participation in worldview education options varies significantly. In 2019, most students (86.1%) in grades 1 to 6 participated in Lutheran lessons, whereas 2.5% participated in Islam lessons and 1.6% in Orthodox lessons. Of all the students, 8.5% participated in Ethics (Vipunen, 2019).

According to the general aims of Finnish religious education, the task is to familiarize students with their own religions and with Finnish religious traditions. In addition, students must also learn about other worldviews. Religious education introduces students to notions of ethical responsibility to help them understand the ethical dimension of religion. Environmental ethical questions are included in the curricula. The curriculum in religious education is divided into three sections: (1) the student's relationship with their own religion, (2) the world of religions, and (3) the good life. In the third section 'the good life', the content in the 1st and 2nd grades refers to the human–nature relationship (FNBE, 2016). In a study of basic education textbooks in Lutheran religious education, Aarnio-Linnanvuori (2013) shows that the human relationship with non-human nature is more widely and profoundly included in textbooks for the seventh to ninth grades, although it is mentioned in the textbooks for the first to sixth grades. The textbooks for first to fourth grades rarely discuss the human–nature relationship, and if so, solely from their own religious perspective. Such mentions are normative and occur mainly in connection with the Creation. To some extent, aesthetics and emotions towards nature are talked about. However, environmental problems and pro-environmental behaviour are rarely considered, and could according to Aarnio-Linnanvuori, be introduced earlier.

8.5 Materials and Method

To uncover how school textbooks make sense of the human–nature relationship, we examined textbook passages that concern this association. We consider schoolbooks to be discursive products that represent mainly objective and useful information (Loftsdóttir, 2010), thereby influencing society. In Laclau and Mouffe's (1985/2001) version of discourse analysis, the Earth should not be seen as a reality that needs to be uncovered to be understood. Instead, they consider that language as well as action construct objectivity. For this study, we analysed how objective knowledge is created regarding human–nature relations in grades 1 and 2 (ages 6 to 8) of basic education. Lutheran and Orthodox religious education follow the same core curriculum and share many traits on the level of a subject curriculum.

The books analysed include 9 textbooks and 7 accompanying assignment books on Lutheran and Orthodox religion published during the period 2015–2022 (see Appendix). The book series, which includes *Himmel och jord* (La1; La2), *Hjärtat* (Lb1; Lb2), *Tähti* (Lc1; Lc2) in Lutheran education, and *Sofian elämä* (Oa1–2) and *Aksios* (Ob1; Ob2) in Orthodox education are referred to in the analysis section through the abbreviations in brackets. Assignment books are referred to through the letter A (e.g., La1A). These books have been used in education during a curriculum that has been in force since 2016 (FNBE, 2016). First, we collected all human–nature relationship references in the texts into the software program NVivo. We found mentions of physical nature, plants and animals, and paid attention to how humans in the text or the addressed reader were framed in relation to these phenomena. We then organized the articulations in the text according to how they were framed. The next step was analysing the categories to find discourses. The analysis was typologically directed (Hatch, 2002) by de Groot and van den Born's (2007) metaphors. Of particular interest was the position and power of humans in relation to the non-human world and whether a change in human–nature relationships in the light of sustainability crises was considered. The discursive patterns in both the Lutheran and Orthodox religious textbooks coincided profoundly and are therefore presented together in what follows.

8.6 Analysis and Results

Three discourses related to de Groot and van den Born's (2007) metaphors on the human–nature relationship dominated in the texts: *Humans as separated from non-human nature*, *Humans as superior to non-human nature,* and *Humans as stewards and caretakers of non-human nature.* A fourth discourse emerged which was relevant to how sustainability crises were considered in the texts: *The Almighty as saviour and guarantor in times of existential crisis.*

8.6.1 Humans as Separated from Non-human Nature

From the very start, all textbooks make a clear distinction between humans and the non-human world. This differentiation is most visible in creation mythologies, which are significant in how the books describe the world to students. A clear dualism between human and non-human nature sets humans distinctively apart from the non-human world as God's final creation. Two of the first-year textbooks, *Hjärtat* and *Aksios*, initially describe the creation story only through images. *Hjärtat* depicts the creation of the non-human world in four smaller images on one page and a separate fifth image that includes humans in a larger one-page-sized image. Separation is explicit in *Aksios* in a dialogue were a student asks whether the teacher had forgotten to tell about the origin of humans in the Bible story: "We will examine the creation of humans quite separately, the teacher Onni reassures" (Ob2, p. 86). Alongside yet separate from humans and non-human nature is the spiritual realm. Apart from the concepts of God and heaven, spirits such as angels are introduced, for instance in relation to God creating "the world, nature, humans and angels" (Ob11, p. 79).

The separation between humans and non-human nature appears strong overall, and they are given distinct roles, which we will examine further below. Only one explicit statement of the world being common to all is found in the books: "God has created the world for everyone" (Lc1, p. 85), However, there is no clear statement of humans being part of nature. Only two of the religion textbooks, *Himmel och jord* and *Sofian elämä*, relate the creation theory to a scientific worldview where humans' interconnectedness to non-human nature is fundamental. *Himmel och jord* suggests that science cannot explain why planets were formed, and religions therefore have their own stories about the origins of Earth. Therefore, the texts do not include a visible discourse of interconnectedness that would reflect the metaphor of *Partner* or *Partnership*.

8.6.2 Humans as Superior to Non-human Nature

Humans' separation from the non-human world is connected to a hierarchical positioning that is most noticeable as a discourse of superiority. This discourse largely coincides with de Groot and van den Born's (2007) metaphor of *Master* and emerges in several ways. It reflects both notions of humans being superior and rulers of non-human nature, and conversely it suggests that non-human nature is of lesser value and is a resource for humans.

The Orthodox textbook *Aksios* states that the world was created for humans: "God created human beings last, when everything was ready for them"[2] (Ob2, p. 86). From this, a view emerges that non-human nature is inferior and has a lesser, secondary value. In contrast, humans are seen as being given a central position in the world. Humans' superiority and power over nature is made clear: "He made humans the ruler and caretaker of the rest of the world" (ibid, p. 87). The text in the same book emphasizes the superiority of humans by alluding to humans' divine character and likeness to God in a chapter heading: "God created humans in his own image and likeness" (ibid, p. 87). Through this statement, non-human nature also

[2] The concepts "ihminen" in Finnish and "människan" in Swedish are used in the curriculum, and translated here as "humans". A corresponding term to the English concept of "Man" does not exist in these languages. Questions of gender in relation to the human–nature relationship are exempted from our analysis.

emerges as less divine than humans, and more distant from the spiritual realm. The Lutheran textbook similarly articulates Man's divine nature: "God wanted that there would be someone on Earth who would take care of his creation and enjoy it. Someone who was like him" (La1, p. 16). God has given the non-human world to humans as a gift and as their possession. This understanding of the human–nature relationship having possession over the non-human world is recurrent in all the books we analysed.

The non-human world is contrasted with humans who can "think and feel" (La1, p. 16), and are described as creative and reflective agents with free will and conscience:

> Humans are capable of creating new things and giving birth to new life. They can think and learn new things. They possess a free will. They are able to choose between good and evil, right and wrong. Their choices are guided by their conscience. (Ob2, p. 88)

The above quotes indicate that the non-human world, including animals, is cognitively and morally on a lower level, and of lesser value than humans, not having the capability of thinking and feeling, and not possessing a free will. This contrasts, moreover, with humans' goodness and closeness to God, a point which is made explicitly: "According to the Bible, humans created by God were similar to God, and were good and eternal" (Ob2, p. 88). The likeness to God is highlighted in humans having an immortal soul, a feature that is not mentioned in connection with animals or other non-human life: "When people die, their body is buried and it will return to dust; the Soul, on the other hand, remains alive" (Oa1–2, p. 45). These books specifically underline that humans can choose between good and bad but also that they can be morally weak and disobey God's will. This is explained through the Fall, when humans lost their innate goodness and were separated from God and the Divine: "Man is no longer like God because he did not follow God's instructions" (Ob2, p.88).

Non-human nature as a resource and utility for humans is strongly present in the books, for instance in the following box information that highlights key knowledge: "God's blessing means that he gave plants, trees and animals the task of flowering, producing seeds and fruit, and multiplying" (Ob2, p. 84). A distant and resource-oriented relation is also expressed by viewing non-human nature as a source of enjoyment: "we may eat tasty apples and enjoy the colourful autumn leaves rustling beneath our feet. We can climb trees, swim in the sea and run along paths in the forest. The cows give us milk, the hens give us eggs and the bees give us honey" (La1, p. 20).

8.6.3 Humans as Stewards and Caretakers of Non-human Nature

Intertwined with the discourse of mastership, the textbooks express a role for humans as stewards and caretakers, coinciding with de Groot and van den Born's (2007) metaphor of the *Steward*. This discourse is to a varying degree present in all books, even if the overarching discourse of caretaking concerns care and respect for fellow human beings rather than nature. It is particularly visible in the story of the good shepherd, which is included in all the book series. Jesus takes care of humans as a good shepherd would his flock – everyone is equal and loved – although this also reflects the relation of leadership and stewardship. As stewards, humans have a sense of care and concern towards non-human nature, but the relation is tempered by a hierarchical understanding, intertwined with the notions of power and responsibility over non-human nature. The following excerpt recognizes humans' dependence on non-human nature, but also highlights caretaking as beneficial for humans.

> We have been given the responsibility for taking care of the earth and everything that is here. Animals also have the right to wellbeing. If we take care of nature, humans also have greater wellbeing. (La1, p. 17)

In one book, a *Partner* discourse is visible through the notion of living in harmony together: "We should live together and respect nature" (Lc1, p. 85). However, this discourse is rarely found in the books. The same textbook also expresses an attitude of gratitude towards nature: "Thank you for the spring, sun, plants and animals" (ibid, p. 89). This quote expresses a need for humans to appreciate and feel thankful for non-human nature, which is visible in other articulations. It could be considered part of a discourse of gratefulness in relation to non-human nature as well as seeing it as intrinsically beautiful and valuable. However, it emerges in the texts as a sub-discourse, intertwined with seeing non-human nature as a gift to humans as masters, and with the responsibility of humans to take care of its beauty and treasures as stewards.

References to humans' inflictions on non-human nature and the protection of Earth from negative influences of human interference are present in the texts. *Himmel och jord* raises the question of how nature can withstand the waste that humans create:

> When God had finished creating, he rested [...]. Nature also needs to rest. When new things are manufactured, it wears nature down. It also becomes hard for nature to cope with all the rubbish. How can we help nature cope? (La1, p. 18)

Similar to *Hjärtat* and *Tähti*, *Himmel och jord* also raises questions for students about how to protect nature. The focus is on what students can do as individuals in everyday life through environmentally friendly deeds, keeping living spaces and the environment clean and tidy, and returning non-human objects back to their original places after studying them. However, the books seldom question the humans' role as good caretakers: "At the beginning of the new church year on the first of September, we pray to God for the Creation. Then we also consider whether we have respected God's work of creation by taking good care of nature" (Oa1–2, p. 120). This is the only place that *Sofian elämä* introduces a critical perspective. However, pondering this question may open a critical discussion in the classroom.

8.6.4 *The Almighty as Saviour and Guarantor in Times of Existential Crisis*

The books promote a fundamental understanding of non-human nature as beneficial for humans. In the creation story, the world is introduced as a fruitful and beautiful garden: "The earth will receive plants, God said. And the earth was filled with trees, flowers, and grass. The trees were full of good fruit and all kinds of berries grew on the earth" (La1, p. 14). Illustrations include flowers, birds, pets and domestic animals, emphasizing a harmless, benign view of nature. Nature is also described as secure and livable: "God created the world as a good and safe place in which to live" (Oa1–2, p. 43), and it is given to humans "as a favourable place" (Ob2, p. 86).

Hence, there is the notion of God having created a pleasant environment for humans, but also having ultimate control, serving as a protector should crisis or need occur. The textbooks present diverse ways through which God, his incarnation Jesus, or angels can affect the course of the world. In this way, a discourse emerges on how the supernatural safeguards humans' wellbeing and protects from evil: "I'm taken care of. The Heavenly father has sent an angel to walk with me" (Lc, p. 38). "Angels monitor the events on earth and fight against evil" (Oa1–2, p. 19). In the parable of the good shepherd, it is stated that Jesus protects humans even in instances of wrongdoing: "Jesus wished to say that he always takes care of people. He does so even if a person sometimes does wrong by straying from his flock" (Ob1, p. 135). In this excerpt God's forgiving nature and his power to place people on the right path, even in the case of wrongdoing, envisions a secure future and a guarantee of safety. Nature is expressed as fearful or as a threat to humans in the books, mainly in cases when humans have not followed God's will. Hence, in the story of Jonah, God created a storm and a huge fish swallows Jonah because he had been disobedient to God's will. In addition, the supernatural emerges as a saviour during a catastrophe. The supernatural has power over the climate and earth and can in case of an emergency perform miracles: the following story is about Jesus falling asleep during a boat trip with his disciples:

Matthew: Master wake up!
Andrew: Can't you see that soon we will all drown?
Jesus: Wind, quiet down and abate.
Storyteller: And so the wind calmed down and the waves subsided.
Jesus: Why were you afraid? Don't you believe I will take care of you?
Storyteller: The disciples wondered and whispered to each other.
Matthew: Who is this man? Even the wind and the waves obey him.
Storyteller: The boat continued sailing and the disciples reached the opposite shore safely. (Lc2 p. 68)

This discourse is also apparent in another Lutheran textbook in the story of Noah's ark when God calls Noah to leave the ark, and promises never again to destroy the earth:

God promised Noah never to wipe out the earth again. He made a colourful rainbow in the sky and said: 'This is a symbol of the alliance I make with all humans. Every time I see a rainbow, I will remember our covenant and floods will no longer devastate the Earth. (Hjärtat1, p. 73–74)

In this excerpt, the special alliance between humans and God is underlined, and through this union, a guarantee is given that the world will not be destroyed. This might be interpreted to mean that catastrophes such as a climate crisis will not and cannot occur. Not to worry is emphasized: 'God takes care of everything – the birds in the sky, the flowers in the meadow and people. You don't need to worry about anything, because God the father knows what you need" (La1, p. 79). The notion that God ensures the wellbeing of humans above all other parts of nature is explicit:

He [Jesus] said: Do not worry about food and clothing. God makes sure the birds are fed. Look at the flowers; can you have clothes that are more beautiful? Just as God takes care of the birds and the flowers, so too he takes care of you. Indeed, he cares about you even more. (La2, p. 39)

The Orthodox textbooks introduce bad spirits as sources of evil. In connection to the creation myth, some spirits turned their back on God, and became bad spirits: "Evil spirits oppose God's work. They try to entice people to do evil deeds. As a result of the work of evil spirits, people may get into accidents or start wars. People can fight evil spirits by praying and doing good deeds" (Oa1–2, p. 43). Even if the book mentions praying and good actions to counter bad spirits, these evil energies may be interpreted as sources of threats and crises to humans – a perspective that consequently may absolve humans from sole responsibility in, for example, the case of natural disasters or climate change.

In a section on children's puzzling questions a question is raised in *Tähti*: Why doesn't God protect humans from floods and storms? A Church mouse answers:

Church mouse: The priest Pirkko has said that God does not plan such disasters. God does not always prevent them from happening. It may seem strange. Still, we can ask God to protect us and our kin.

Saara: Maybe it's good that we don't know what will happen tomorrow. (Lc2, p. 89)

Saara's answer that it is good that we do not know what the future entails can be helpful in protecting children from worries, particularly at an early age. Similarly, the presentation of the world as a safe and good place, and the Almighty as an ultimate protector may support children's' feelings of faith and trust in the world and the future. However, the articulation above as well as the presentation of the supernatural as ultimate protector can also support a passive and naive attitude towards the surrounding world, which may have a blindfold effect on how children direct themselves in relation to developments that humans need to confront and take responsibility for.

8.7 Discussion

Under the heading of the underlying values of basic education, the Finnish Core Curriculum 2014 explicitly states, "humans are part of nature and are totally dependent on viable ecosystems" (FNBE, 2016, p. 14). Hence, it appears contradictory that the textbooks in Lutheran and Orthodox religion formed a distinct view of separation and dualism between humans and non-human nature. As Martin (2015) also argues, our results raise a concern of how dominant human-centred positions in religious education can encourage students to recognize the interconnectedness of all life. In the analysed textbooks, a worldview of interconnectedness was almost non-existent. This coincides with the outdated dominant human–nature relation in Western and Christian thought that causes concern from a sustainability perspective. The human–nature relationship was presented to students through a discourse of superiority, and did not visibly reflect the metaphors of *Partner* or *Partnership* (de Groot and van den Born, 2007). Similarly, the steward and caretaker discourse reflected a limited view of existential interconnectedness and nurtured a hierarchical understanding of human relationships with non-human nature. Moreover, discourses that would support a change and a re-imagining of the human–nature relationship that sustainability research and policies call for were not visible in the textbooks. Instead, a discourse representing the Almighty as a saviour and guarantor in times a crisis emerged that may not engage students in developing their relation to non-human nature. Like Martin (2015, p. 34), we conclude that "a gap remains between our appreciation of the reality of ecological devastation and our work in religious education toward changing the way we think and live".

As Rissanen et al. (2023) argue, the sustainability crisis challenges worldview education scholars to step out of the margins and participate in the transformation of the worldview basis of education. In line with calls from policy and research, this rethinking needs to take place among both educators and students, and requires a multidisciplinary and an inclusive effort. The need to reframe the human–nature relationship in worldview education extends to questions of theology. Without rethinking humans' self-understanding on a religious institutional level, it is difficult for both teachers and students to discuss and rethink the human–nature relationship in the classroom. As Sveungsson (2018) argues, living religions need to problematize the repressive or destructive features of religious tradition and at the same time discern and encourage its benign sides. Efforts to revise Christian theology in ways that embrace interrelatedness and an attitude of care in the human–nature relationship are now apparent both internationally and in Finland (Jenkins et al., 2018; Koij, 2021; Nantsou & Asproulis, 2021).

We see a substantial need for the re-examination and development of Lutheran and Orthodox religious education textbooks to provide more opportunities to learn about various human–nature relationships, to reflect on their differences, and re-imagine new ones (see also Aarnio-Linnanvuori, 2013). A general updating of Finnish worldview education and syllabi to include more integrative approaches (e.g., Kavonius, 2021) seems crucial to support this development. We challenge

Finnish worldview education and textbook writers to make stronger multidisciplinary efforts. This is in line with Aarnio-Linnanvuori (2013) and Martin (2015), who also call for the development of scientific perspectives in worldview teaching materials. Useful in this respect is Tomlinson (2019), who describes a set of practices for religious education that integrates ecology in Christian education, including teaching materials of narratives derived from Christian traditions as well as scientifically based stories about the natural world.

There is also a need to encourage students' critical reflection and agencies with respect to the sustainability crisis. Students in the first grades may not engage in advanced critical reflection, but teaching materials can encourage reflection and dialogue on humans' position in the world. This can be done in several ways, namely by engaging thoughts and emotions that are grounded in children's own experiences, and developing a situatedness that includes pluralistic worldviews and both local and global perspectives (e.g., Kvamme, 2020; Aarnio-Linnanvuori, 2013; Zilliacus & Wolff, 2021). Education on religion and the environment could include texts that develop literacy in the environmental implications of religions and different environmental commitments. This would help students to articulate and refine their own environmental advocacy (O'Brien, 2014). Finally, Altmeyer (2021) sees the idea of religious stewardship as a relationship where religious education could present different motivations for engaging in the role of stewardship.

Religious education could emphasize multiple and optional actions and solutions, and encourage pro-environmental deeds rather than focusing on alarming and distressing matters (Aarnio-Linnanvuori, 2013). Bouma-Prediger (2016) develops perspectives on religious pedagogy through ecological virtue ethics. Virtues such as ecological attentiveness as well as equity as a concept that extends to non-human creatures could be introduced into teaching materials such as stories and biographies about earth keeping. Teaching materials that encourage students to engage in embodied and emotional experiences are other foundations to view relationships with nature in a holistic way. For instance, *Wild pedagogy* emphasizes the need to give space to experiences in education and strives for a deeper connection with the non-human world (Blenkinsop et al., 2022). In this view, the natural world is a co-teacher that is vibrant, active and worth listening to. In the framework of a relational ontology, Kuchta (2022) discusses the epistemological possibilities of love and of relearning love of the land.

Our study shows a problematic situation regarding religious education textbooks that reflects larger problematic discourses within religions. Where human–nature relationships are concerned, religions are ambiguous – theological interpretations vary and there are various perspectives in Christianity as well as in other worldviews. This plurality of views and the possibility for re-interpretation open up the potential for new understandings, revisions and developments of the human–nature relationship. This also provides impetus to improve worldview education and create better teaching materials that encourage a positive view of sustainability in the future.

Appendix: Textbooks

Orthodox textbook series Aksios. Publisher: Opetushallitus

Aikonen, R. & Havu-Nuutinen, S. (2017). Textbook (Ob1) and Assignment book, grade 1 (Ob1A).
Aikonen, R. & Havu-Nuutinen, S. (2019). Textbook (Ob2) and Assignment book, grade 2 (Ob2A).

Orthodox textbook series Sofian elämä. Publisher: Opetushallitus

Kantonen, M. & Tajakka, T. (2009). Assignment book, grades 1 and 2 (Oa1–2A).
Kantonen, M. & Tajakka, T. (2021). Textbook, grades 1 and 2 (Oa1–2).

Lutheran textbook series Himmel och jord. Publisher: Fontana media

Erikson-Blomfelt G., Sjö-Särs, I., & Fontana Media. (2016). Textbooks, grades 1 (La1) and 2 (La2).

Lutheran textbook series Tähti. Publisher: Otava.

Kirkkopelto, K., Niittynen, A., Similä, M., & Vidgren, K. (2008). Textbook, grade 1 (Lc1).
Kirkkopelto, K., Niittynen, A., Similä, M., & Vidgren, K. (2009). Assignment book, grade 1 (Lc1A).
Niittynen, A., Similä, M., & Vidgrén, K. (2009). Textbook (Lc2) and Assignment book, grade 2 (Lc2A).

Lutheran textbook series Hjärtat. Transl. from Finnish by C. Heinonen. Publisher: Otava.

Niittynen, A., Similä, M., & Vidgrén, K. (2015). Textbook, grade 1 (Lb1).
Niittynen, A. & Similä, M. (2015). Assignment book, grade 1 (Lb1A).
Niittynen, A., Similä, M., & Vidgrén, K. (2016). Textbook (Lb2) and Assignment book (Lb2A), grade 2.

References

Aarnio-Linnanvuori, E. (2013). Environmental issues in Finnish school textbooks on religious education and ethics. *Nordidactica, 1*, 131–157.

Altmeyer, S. (2021). Religious education for ecological sustainability: An initial reality check using the example of everyday decision-making. *Journal of Religious Education, 69*(1), 57–74.

Attfield, R. (1994). *Environmental philosophy: Principals and prospects*. Avebury.

Berry, T. (1988). *The dream of the earth*. Sierra Club Books.

Blenkinsop, S., Morse, M.& Jickling, B. (2022). Wild pedagogies: Opportunities and challenges for practice. In M. Paulsen, J. Jagodzinski, & S. M. Hawke (Eds.), *Pedagogy in the anthropocene: Re-wilding education for a new earth* (pp. 33–51). Palgrave Macmillan.

Bouma-Prediger, S. (2016). What kind of person would do something like that? A Christian ecological virtue ethic. *International Journal of Christianity & Education, 20*(1), 20–31.

Callicott, J. B. (2013). *Thinking like a planet: The land ethic and the earth ethic*. Oxford University Press.

de Groot, M., & van den Born, R. J. G. (2007). Humans, nature and God: Exploring images of their interrelationships in Victoria, Canada. *Worldviews, 11*(3), 324–351.

Finnish National Board of Education (FNBE). (2016). *National core curriculum for basic education*. FNBE.

Goetz, S. (2011). *Freedom, teleology, and evil: Religion and spirituality across cultures*. Springer.

Haraway, D. (2018). Staying with the trouble for multispecies environmental justice. *Dialogues in Human Geography, 8*(1), 102–105.

Hatch, J. A. (2002). *Doing qualitative research in educational settings*. SUNY.

Jenkins, W., Berry, E., & Kreider, L. B. (2018). Religion and climate change. *Annual Review of Environment and Resources, 43*(1), 85–108.

Kavonius, M. (2021). *Young people's perceptions of the significance of worldview education in the changing Finnish society*. Unpublished doctoral dissertation. University of Helsinki.

Koij, S. (2021). *Resiliensteologi: En studie av den kristna traditionens hållbara förändring* [Resilience theology: A study on the sustainable change of the Christian tradition]. Unpublished doctoral dissertation. Åbo Akademi University.

Kuchta, E. C. (2022). The epistemological possibilities of love: Relearning the love of land. In M. Paulsen, J. Jagodzinski, & S. M. Hawke (Eds.), *Pedagogy in the Anthropocene: Re-wilding education for a new earth* (pp. 53–69). Palgrave.

Kvamme, O. A. (2017). The significance of context: Moral education and religious education facing the challenge of sustainability. *Discourse and Communication for Sustainable Education, 8*(2), 24–37.

Kvamme, O. A. (2020). Situating moral education in a globalized world: Environmental ethical values and student experiences. In T. Strand (Ed.), *Rethinking ethical-political eeducation* (pp. 45–65). Springer.

Laclau, E., & Mouffe, C. (1985/2001). *Hegemony and socialist strategy: Towards a radical democratic politics*. Verso.

Loftsdóttir, K. (2010). Encountering others in Icelandic textbooks: Imperialism and racial diversity in the era of nationalism. In T. Helgason & S. Lässig (Eds.), *Opening the mind or drawing boundaries? History texts in Nordic schools* (pp. 81–98). V&R unipress.

Martin, M. K. (2015). The cry of earth and conflict with human cultures: A reflection for Christian religious educators. *Journal of Religious Education, 63*(1), 25–36.

Mayer, F. S. (2019). *Transforming psychological worldviews to confront climate change: A clearer vision, a different path*. University of California Press.

Nantsou, T., & Asproulis, N. (Eds.). (2021). *The orthodox church addresses the climate crisis*. WWF Greece & Volos Academy Publications.

O'Brien, K. J. (2014). Balancing critique and commitment: A synthetic approach to teaching religion and the environment. *Teaching Theology & Religion, 17*(3), 189–202.

Paulsen, M., Jagodzinski, J., & Hawke, S. M. (Eds.). (2022). *Pedagogy in the Anthropocene: Re-wilding education for a new earth*. Palgrave.

Riechers, M., Balázsi, Á., García-Llorente, M., & Loos, J. (2021). Human–nature connectedness as a leverage point. *Ecosystems and People, 17*(1), 215–221.

Rissanen, I., Aarnio-Linnanvuori, E., & Mansikka-aho, A. (2023). Worldview transformation in and through education: Mapping the nexus of climate education and worldview education. In *Religion and worldviews in education* (pp. 194–206). Routledge.

Salmenkivi, E., Kasa, T., Putkonen, N. E., & Kallioniemi, A. (2022). Human rights and children's rights in worldview education in Finland. *Human Rights Education Review, 5*(1), 47–69.

Skilbeck, M. (2021). *Loving and studying nature: Celebrating the earth through history, culture and education*. Springer.

Stevenson, R. B., Brody, M., Dillon, J., & Wals, A. E. J. (2013). Introduction. In R. B. Stevenson, M. Brody, J. Dillon, & A. E. J. Wals (Eds.), *International handbook of research on environmental education* (pp. 1–7). Routledge.

Svenungsson, J. (2018). J. Drömmen om en annan värld: Det bibliska arvets utopiska kraft. In M. Lindman & J. Ahlskog (Eds.), *Rum för eftertanke: en antologi om att utmana det invanda* (pp. 45–59). Folkets bildningsförbund.

Tomlinson, J. (2019). Ecological religious education: New possibilities for educational practice. *Journal of Religious Education, 67*(3), 185–202.
UNESCO. (2021). *Re-imagining our futures together: A new social contract for education*. https://unesdoc.unesco.org/ark:/48223/pf0000379707
Vipunen. (2019). Kieli ja muut ainevalinnat [Language and other subject choices]. https://vipunen.fi/fi-fi/perus/Sivut/Kieli%2D%2Dja-muut-ainevalinnat.aspx
von Wright, G.-H. (1994). *Att förstå sin framtid: Tanke och förkunnelse och andra försök 1945–1994* [On understanding one's own time: Thought, speech and other attempts 1945–1994]. Bonniers.
Wilson, E. O. (1993). Biophilia and the conservation ethic. In S. R. Kellert & E. O. Wilson (Eds.), *The biophilia hypothesis* (pp. 31–41). Island Press.
Wolff, L.-A. (2011). *Nature and sustainability: An educational study with Rousseau and Foucault*. Lambert Academic.
Wolff, L.-A. (2022). Ethics of sustainability education. In S. Idowu, R. Schmidpeter, N. Capaldi, L. Zu, M. del Baldo, & R. Abreu (Eds.), *Encyclopedia of sustainability management*. Springer. https://doi.org/10.1007/978-3-030-02006-4_1139-1
Zilliacus, H., & Wolff, L. A. (2021). Climate change and worldview transformation in Finnish education policy. *Oxford Research Encyclopedias Education*. https://doi.org/10.1093/acrefore/9780190264093.013.1676
Zimmermann, J. (2015). Being human, becoming human: Christian humanism as a foundation of Western culture. In D. Melé & M. Schlag (Eds.), *Humanism in economics and business: Perspectives of the Catholic social tradition* (pp. 49–67). Springer.

Harriet Zilliacus is a University Lecturer in educational sciences at University of Helsinki, and holds the title of Docent at the Faculty. Her research focuses on sustainability education, and questions of worldview transformation through education. Among her research interests are also intercultural education, and questions of diversity and inclusion in education.

Lili-Ann Wolff is an Associate Professor and Senior Lecturer at the Faculty of Educational Sciences at the University of Helsinki. Based on degrees in education, science and philosophy, her research includes sustainability education and education in the Anthropocene from transdisciplinary and critical views combining philosophy of education, and science didactics.

Pia Mikander is a University Lecturer in history and social studies didactics at the University of Helsinki. Since her PhD in 2016, *Westerners and others in Finnish school textbooks*, her research interests have focused on questions around democracy, active citizenship and anti-racism in education, particularly within the subject of social studies.

Arto Kallioniemi is Professor of religious education at the University of Helsinki. His research interests are worldview education and its profile in the future. Furthermore, human rights education is in the focus of his current research. He also holds as UNESCO Chair on values, dialogue and human rights.

Chapter 9
Indigenous Spirituality with a Focus on Maya Religion and Ritual Practices in Connection with Education on Sustainable Development in RE Classrooms

Bodil Liljefors Persson

9.1 Introduction

The theme religion, education and sustainability is urgent to explore and it puts an emphasis on how encounters between cultures, religious traditions and societal changes prompts education, action and innovations to meet global challenges like poverty, social inequalities and climate crises. There is a need to have knowledge of the historical processes of change of religions and to understand the meaning of religion in contemporary society. This is illustrated by the fact that religious traditions still are important in the majority of people's everyday lives in the world. Also, many cultural contexts recognise the existence of a spiritual dimension of the natural environment and spirituality is often connected to places in the landscape, like mountains, lakes and caves. These beliefs are found in many ancient traditions as well as in contemporary religions, sometimes they are no longer practiced, or have undergone changes, but at the same time new interpretations are currently developing and framed in new spiritual contexts. There is also an increased interest in discussing man's relationship with nature, and of global values, global ethics and responsibilities in regard to both local landscapes and global environment. This is not the least seen formulated in the Agenda 2030 for Sustainable Development that was adopted by all United Nations Member States in 2015. It is therefore of great interest to explore the relationship and interaction between religion, education and sustainability for both religious studies research and teaching in religious studies and religious education.

The purpose with this contribution is to present Maya religion as an empirical example that may be used in various educational contexts when connecting

B. Liljefors Persson (✉)
Malmö University, Malmö, Sweden
e-mail: bodil.liljefors@mau.se

S. Windsor, O. Franck (eds.), *Intersections of Religion, Education, and a Sustainable World*, https://doi.org/10.1007/978-3-031-81809-7_9

indigenous religions and sustainability, and to inspire teachers and student teachers to develop approaches that combine indigenous religions with ecology, spirituality, sustainability and global ethics in various educational contexts and at various educational levels.

In the first part of the chapter the background of the concept of Indigenous religions will be presented and a brief overview of some common components in Indigenous religions in general. Thereafter, a second part will highlight some central themes in Maya cosmology, mythology and rituals with the viability of Maya views on nature, agriculture and ecology. Local knowledge of Indigenous religions and spirituality, exemplified by the case of Maya cosmology and ritual practices especially in relation to agriculture and to ritual healing, will be related to, and deepen the understanding of processes of change, permanence and thus of sustainability. Maya religious traditions still today communicate meaning and spirituality in the Maya areas in Mexico, Belize and Guatemala. Finally, the third part of this contribution will focus on Education for Sustainable Development and on combining powerful knowledge and controversial issues in Religious Education. Here the objective is to integrate questions of sustainability and ecology in Religious Education at various levels and to encourage future teachers in Religious Education to develop their students' knowledge on the relations between religion(s) and sustainability, and to suggest how to teach sustainability education in the RE classroom.

9.2 Indigenous Religions – Concept and Content

When talking about the religions of Indigenous peoples, it is imperative to emphasize that it is precisely about the religions of different Indigenous peoples around the world, which also may be characterized by diversity within the respective tradition. In the past it was common to use the concept Nature religions, non-scripture religions, or even primitive religions, when talking about the different Indigenous religions in the world. Nowadays, these designations are not used often, for slightly different reasons. Nature religions was a definition that was based on the idea that the different groups were considered living more in close harmony with nature than other religious groups. This concept may still be use in contexts where there is a strong emphasis on the close connection with humans and nature and with environmental questions. Primitive religions was a term used partly because it referred to what was original and thus of older chronology, and because it was associated with simple and not so advanced societies. The Non-scripture religions included, for example, the notion that there were no written sources or sacred texts among these traditions (Andersson & Liljefors Persson, 2024; Liljefors Persson, 2017). All these explanations show how those concepts are outdated and are therefore used less today. Nowadays Indigenous peoples in the world do not only live in the countryside and thus closer to nature, nor are they completely unaffected by the surrounding global world or illiterate.

The lives and religions of indigenous peoples are no less complex than the monotheistic religions which in earlier research in comparative religion and according to the paradigm of evolution, were considered to constitute the highest form of the evolution of religion. This pattern of hierarchy developed in the late nineteenth century in the context where Darwin made his discoveries and the theories of the survival of the fittest and the evolutionary scientific paradigm became dominant (Andersson & Liljefors Persson, 2024; Cusack, 2016). In the academic studies of religion during the twentieth century there were debates where researchers considered the monotheistic religions being too anthropocentrically orientated and not attentive to the meaning and importance of nature. Other researchers have instead emphasised that the religions in Asia as well as the Indigenous religions are more peaceful and stress the importance of keeping a balance between humans and nature. (Taylor, 2011; Cotter & Robertson, 2016; Mulk & Bayliss Smith, 2007). There has been an emphasis on bringing back the importance of the human-nature-relationship within Judaism, Christianity and Islam. Also, there has been suggested a difference between green religion, which promotes that "environmentally friendly behaviour is a religious obligation", which sometimes is labelled religious environmentalism, and dark green religion, in which "nature is sacred, has intrinsic value, and is therefore due reverent care" (Taylor, 2011, p. 19).

With influences from postcolonial theory-building that strives to go beyond an approach characterized by an "us and them" thinking, it becomes important to apply/utilise the names and concepts that the ethnic groups use about themselves (Hall, 1992). Therefore, it is more common today in research and in international bodies/institutions, such as the United Nations, to employ the term Indigenous religions in English or Indigenas in Spanish. When talking about the various religions of indigenous peoples in general, and when one wants to highlight more general knowledge and information of various kinds, it is thus better to employ the term Indigenous religions than the above-mentioned and previously common concepts. However, when speaking of a particular Indigenous population in a certain part of the world more generally, it is preferable to rather use names that the ethnic group uses about themself, e.g., Sami, Maori, Yoruba, Maya or Mi'kmaq. Sometimes you can also add a further specification for example if it is a larger group that speaks different languages, such as Yucatec Maya or Swedish Sami.

Common to many of the Indigenous religions is a central belief in the notions of sacred landscape and sacred ecology. Many of the varied groups consider that both the natural landscape and places created by humans are loaded with spiritual meanings and beings. Indigenous groups often emphasise that there is a close connection between humans and nature and that all humans need to take responsibility for maintaining a healthy balance with nature and not exploiting the land, and that this is often maintained by performing various ritual practices. It is also common to emphasize the ways in which indigenous peoples try to harmonize their everyday life with the rythms and cycles found in nature, which will be evident below where the focus will be on the Maya culture and religion.

9.3 Maya Cosmology and Ritual Practices

In the Maya area of Mesoamerica, around 8 million Maya descendants live today in various ethnic groups who speak 30 different Mayan languages.[1] The most widely spoken language is Yucatec Maya, spoken by approximately 800,000 people in the Yucatan Peninsula, the majority of whom are bilingual with Spanish as the official language of Mexico and Guatemala. There is an awareness that it is important to preserve the linguistic and cultural heritage of various minority groups, as well as work actively to allow all ethnic groups to practice their religion and religious practices. Bilingual schools are found all over Mexico and Guatemala, and even daily newspapers can be partially bilingual. Interest in Maya pre-Hispanic history and culture is increasing and a tangible return of traditional beliefs and rituals is noticeable both in cities and in the countryside.

The culture, history and religion of the Maya people is perhaps the most studied of all the cultures in the geographical area of Mesoamerica. We know that various Maya groups have built city-states and inhabited the southern states of Mexico, Guatemala, Belize, Honduras, Costa Rica and El Salvador since about 1500 BC. During the Classical period (around 250 to 900 AD), cities with temples, pyramids, astronomical observatories, ball courts and other buildings where the elite families lived. These cities were surrounded by smaller communities where the peasant population lived and where they cultivated the land and forest. Today, the Maya area is being explored with the help of advanced measuring instruments such as lidar technology (light detection and ranging), which means that new urban areas will be possible to excavate from the dense jungle areas that hide urban settlements from the classical period and older times. The knowledge we have so far about Maya culture and its history and religion will thus develop in the future as new scientific findings are generated by combining and relating the new discoveries with lidar technology with new archival finds, fieldwork and reinterpretations of texts dating from Pre-Hispanic times and onwards.

There are four hieroglyphic manuscripts left from the pre-Hispanic time called Codex Dresden, Codex Paris, Codex Madrid and Grolier Codex (Houston & Inomata, 2009). The manuscripts derive from twelfth to thirteenth centuries and have been named after the places where they are kept in various libraries and museums. The central content in these are calendars and iconographic motifs of gods and mythological figures depicted in different contexts, such as hunting rituals or river floods, painted on paper made from amate, bark from fig trees. These four codices

[1] Within Maya research and in scholarly literature as well as in common speech, it is customary to use Maya as both a singular and plural noun i.e. in Maya people, both in a historical perspective and in present time, and as an adjective, such as in Maya culture, Maya rituals and Maya identity. The only exception is when referring to the various language groups, as in Tzotzil Mayan language, or Quiché Mayan language. And even when referring to these various language groups there is an exception regarding the language on Yucatan where the common use is Yucatec Maya (Sharer, 2009; Castaneda, 2021).

are the oldest known examples of written literature from both the American continents.

The so-called Chilam Balam books, the books of the jaguar prophet, constitutes another central part of the source material for the knowledge we have about Mayan religion and history. They were composed during the early colonial period on the Yucatan Peninsula in Mexico and are written with Latin letters in the Yucatec Maya language. They have received their name after Chilam Balam, the jaguar prophet, who lived and was active in Mani in Yucatan at the beginning of the sixteenth century. Today, about 10 different Chilam Balam books remain, but 18 such books are known. They have been named after the village where they were stored and possibly also compiled, for example, the Chilam Balam book from the village of Chumayel (Bricker, 1981; Bricker & Miram, 2002; Suarez Castro, 2017). These manuscripts are very heterogeneous and contain for example myths, rituals, texts with geopolitical content, historical texts, prophecies, medical prescriptions and remedies, astronomical observations, calendars, biblical material as well as other texts influenced of European thought (Liljefors Persson, 2000).

Along with other early colonial texts, such as the Ritual of the Bacabs (Roys, 1965) and Bishop de Landas Relación de las cosas de Yucatan dating from 1566, (translated and edited by Tozzer, 1941), the Books of Chilam Balam are the most important sources for our knowledge of Maya history and religion from the colonial period in the Yucatan Peninsula. Popol Vuh, The Council's Book, in another known manuscript from the colonial period which contains a creation epic from Guatemala as well as stories about how the first people groups populated Mesoamerica (Tozzer, 1941; Arzápalo, 1987; Christenson, 2007).

9.3.1 Cosmology, Myths and Rituals

What may be called the Maya spiritual or sacred landscape under the classical period encompassed nature in its entirety with mountains, valleys, lakes and rivers as well as the cultural landscape with the cities with temple buildings, ball courts, houses and other buildings. The sacred landscape was considered filled with spiritual power and energy. At certain special points in the world the sacred powers are especially concentrated, for example in caves (chen), mountains (witz) and in the so-called dzonots or cenotes, sinkholes in the limestone foundation. These are considered particularly important for performing rituals and some of these places are important centres for pilgrimages even today (Houston & Inomata, 2009; Liljefors Persson, 2011a, 2014, 2018a, b; Schele & Mathews, 1998; Taube, 2004; Williams-Beck et al., 2012).

In the creation myths found in the various Chilam Balam-books we can read how the Maya imagined the world divided into three parts (Roys, 1967; Craine & Reindorp, 1979). At the top was the sky called Ox-lahun-ti-ku, 13 gods, then followed an intermediate world where the people were mainly thought to live, and finally the lowest part, the underworld called Bolon-ti-ku, nine gods. These three worlds were considered interconnected, and by performing blood sacrifices, ceremonies or ritual dances, the gods, mythical figures and Mayan ancestors were believed to be able move between the three parts (Liljefors Persson, 2006; Houston & Inomata, 2009; Taube, 2004). The part were the humans lived was thought to float around in a primeval sea, sometimes on the back of a caiman, sometimes on the back of a tortoise called Itzam-cab-ain. The world was believed to be oriented according to four cardinal points – east, south, west and north which in the Maya creation stories (Roys, 1967) have been symbolically depicted as a tree with a bird on the top, in different colours representing the cardinal directions: red for east, yellow for south, black for west and white for north. In the middle was an eternal blue-green tree that connected heaven and underworld. This world tree is sometimes called the Tree of Heaven (Wacah Chan) or the First or the green-blue tree (Imix Yax Che).

In iconographic motifs, the world tree is a common symbol. One reason why Christian beliefs were accepted relatively quickly in some groups may be due to the central role of the cross in pre-Hispanic cosmology and that the cross as a symbol had similarities with the depictions of the world tree in the iconography. To this day, the cross has a strong symbolic position. At the entrance to some of the smaller towns and villages in the Yucatan Peninsula we find crosses wrapped in beautiful fabrics or dressed in the traditional dress for women, huipil. Also, in the traditional churches, crosses are often seen wearing huipil (Liljefors Persson, 2018a, b, 2023a, b).

The creation of the world and time – man at the centre of the world.

In the Chilam Balam books of Chumayel and Tizimin there are detailed accounts of how the world was created and divided in the pattern with the world tree and its four cardinal points (Roys, 1967; Edmonson, 1982; Mathews & Garber, 2004; Liljefors Persson, 2000, 2011a, b). These books tell the reader about how important it is with the four cardinal points in connection with rituals, i.e. for instance in preparing the various dishes, plants and herbs when setting the table and the altars used in several of the agrarian rituals which are still carried out to this day in large parts of the Yucatán Peninsula. Usually, these rituals are carried out in a special place in nature or at a corn field. They build an altar with leafy branches in each of the four corners of the table and then set out drink and food offerings like corn beer (atole) and tamales or tortillas that are sacrificed and consumed in a communal meal (Redfield & Villa Rojas, 1962; Gubler & Bolles, 2000; Love, 1986; Liljefors Persson, 2011b, 2018a, b).

The Creation Story in the Chilam Balam Book of Chumayel begins with a declaration that the time to preserve this story is now, so that it is not forgotten when the old gods disappear. The world has been created at least three times, it says, and each time it has been destroyed by some sort of natural disaster: floods, earthquakes or hurricanes (Roys, 1967). We can also read that during the creation itself a certain number of plates with different dishes are prepared, seven, nine or thirteen pieces, as well as bowls of drink, and together they are placed in a particular order. Today when the altar is laid out in the beginning of various agricultural rituals that are still performed in Yucatán these different sequences of the creation become visible during the ceremony. In descriptions from the town of Chan Kom from the 1930s the number of plates and drinking vessels are the same as during the rituals in present day Yucatan. The Cha Chaak rituals are still performed to bring rain to the crops (Redfield & Villa Rojas, 1962; Love, 1986; Gubler, 2005; Liljefors Persson, 2011a). Also, Diego de Landa (Tozzer, 1941) a bishop who lived in the early colonial period in the sixteenth century, mentions food offering rituals being carried out at the New Year (Wayeb), described in a similar way. This basic structure is repeated as a basic pattern in all the agrarian rituals, for example Tup kak and Cha Chaak, which still are practiced in Yucatán. The ritual leader, the hemnoob, is seen as creating a microcosm of the world when the altar is set up. In this microcosm the world is circled by placing branches in each cardinal point around the table as a preparation to represent the macrocosmos pattern of the globe. This worldview is repeated in both myths and rituals from early colonial times and is still central in the belief of what can be called the lived religion or living religious traditions of Yucatán (Tozzer, 1941; Redfield & Villa Rojas, 1962; Love, 1986; Liljefors Persson, 2000, 2020). However, it is not only the rituals that endure through time and space, today the Maya work in agriculture and forestry are noticed as sustainable and ecological, because they have worked with management of the natural resources with a resilient capacity in the same way for centuries (Barrera-Bassols & Toledo, 2005).

9.3.2 *Maya Religion, Resilient Rituals and Sustainability – What and Why in Religious Education?*

The traditional Mayan medical knowledge is based on Mayan cosmology and contact with the spiritual world and nature. There is a belief in a holistic view of the body, where health depends on a balance between body and soul. The natural environment affect everyday life and humans are seen as an integral and interacting part of the cosmos and society. Therefore, a person should maintain a good balance in their relationship with nature, to sustain balance in the relationship between humans and the cosmos.

Male and female shamans – hmenoob and xmenoob – have been the most central of the various religious specialists of Yucatan for centuries. The majority are men and the designation for them, hmen (singular) or hmenoob (plural), can be translated as "doer" or "one who does something". The shamans lead various agrarian rituals that are necessary to be able to live a good, balanced life and conduct good agriculture in interaction with the gods. They are also often hired to cure certain diseases, which, for example, have been caused by "evil winds" (Redfield & Villa Rojas, 1962; Hanks, 1986; Jones, 1989; Gubler & Bolles, 2000; Gubler, 2005, 2017; Hirose Lopez, 2015).

The most common type of ritual performed by a hmen, is the blessing (santiguar), and it is carried out either for preventive purposes or to cure diseases caused by evil winds. A hmen must get the right education in the arts to use their sacred crystals (zaztunoob), to be able to perform divinations and diagnoses, or to foretell the future of one who seeks help from them. There are many examples of medicinal rituals reproduced in the various Chilam Balam books and in the Rituales de los Bacabes from the eighteenth century (Roys, 1965; Hanks, 1990; Liljefors Persson, 2000, 2011b).

The traditional Mayan rituals are rituals loaded with meaning and intricate symbolism. Much has changed over the long term, but certain rituals and spiritual practices carry on old traditions and may be seen as resilient and as socially sustainable as well. It is important to continue to collect and document material about the rituals to preserve them because many of the village shamans are old and many of the young people today do not want to learn the old traditional ways of life. This tension between traditionalists and modernists and secularists is constantly present in Yukatán.

Over the course of history, there have of course been changes both in ritual processes and regarding what is considered meaningful in the rituals that are performed, but we can also see how the world of ideas and imagination is reinterpreted by Christian influences and that new age spirituality and new ritual practices emerges. It is essential, however, to understand that there is still a strong continuity in both Maya religious beliefs and in the ritual practices despite the changes that have taken place.

As a teacher, it can be particularly interesting to highlight Maya religion and culture in teaching to arouse students' interest in understanding different ways of life and of how people practice their religion in various local contexts. In this way, students can be given the opportunity to broaden and deepen their understanding of what religion means in different times and in different geographical locations (Cotter & Robertson, 2016). In this context it might be of interest to discuss questions about both ontology and epistemology and local knowledge. Discussions about the relation between man and nature and about global values as well as global ethics and responsibility could be of particular interest. Also questions and themes relating to ecology and sustainability from the perspective of various indigenous religions, i.e. the Saami or the Maya, and also in a comparative perspective (Liljefors Persson, 2023a, b).

9.4 Education for Sustainable Development and Religious Education in the Classroom

In September 2015, world leaders adopted the declaration on the 2030 Agenda for Sustainable Development with its 17 Sustainable Development Goals and 169 targets. It is a global action plan whose objective is to transform the world and it builds on the Millennium Development Goals and "provides a universal, transformative, ambitious, shared and common vision for all humankind, all religions and cultures, and all creatures on earth." (Agenda 2030; UNEP, p. 4; World Commission on Environment and Development (WCED), 1987). The Agenda 2030 also intended to stimulate action in the five areas of people, planet, prosperity, peace and partnership with critical importance for humanity and the planet as a whole (UNEP, p. 5).

In a 2016 document titled 'Environment, Religion and Culture in the Context of the 2030 Agenda for the Sustainable Development' from the United Nations Environment Programme (UNEP), we learn that over 80% of people globally identify as being religious and it is estimated that over 4300 religious or spiritual communities are found worldwide (UNEP, 2016). It stresses that many global problems are continuing and occur more frequently, such as natural resource degradation and more intense natural disasters, and inequalities in and among countries rises as well as unemployment, particularly for youth, which is a global major concern. Other challenges include violent extremism, terrorism, humanitarian crises and forced displacement of people as these counteract developments that have been made, so work will need to continue with the goals of Agenda 2030 in focus (UNEP, 2016).

The concept of "Going green" is related to the belief of spiritual connectivity between nature and humans, and that religious communities must work towards understanding and having respect for cultural diversity. This also includes various forms of secularism. UNEP has formulated that a shared vision for humanity is at the "heart of building an ethic of global and local citizenship." (UNEP, p. 19). Similar visions and goals for sustainability, human values and global ethics are central in much research and international collaborative work, thus for the interested teacher and student the possibilities to learn are almost endless.

9.5 Religious Education, Powerful Knowledge and Sustainability

From a European perspective, the teaching content appears to be that religious knowledge is defined differently in different countries, but also much exists that is similar in Religious Education in many countries. Many course plans highlight the importance of multiculturalism and religious diversity, and to civil formation (Alberts, 2010; Heimbrock et al., 2001; Jackson, 2004, 2008, 2019; Williams et al., 2008).

Different subjects have their own unique subject content where particular concepts become important for students to learn (Shanahan & Shanahan, 2012). Thus, students need to become subject literate in various subjects, in our case here it is Religious Education (RE). The aim is therefore that students should be able to use their subject knowledge in an adequate way in different contexts in life - they need to become religiously literate, so to speak. Sometimes the term religious competence is also used for precisely this.

There is increasing talk of powerful knowledge and similar concepts, such as core concepts and big ideas. They refer to what is considered central knowledge in the subjects and the concepts are also used in the research literature (Meyer & Land, 2005; Young, 2008, 2014; Jackson, 2019; Liljefors Persson, 2023a, b; Franck & Liljefors Persson, 2023).

Young defines the concept of powerful knowledge as follows:

> Powerful knowledge refers to what the knowledge can do or which intellectual power it gives to those who have access to it. Powerful knowledge provides more reliable explanations and new ways of rethinking and engaging with the world, and it can provide students with a language for engaging in political, moral, and other types of debate (Young, 2008, p. 14).

Powerful knowledge helps students make comparisons, both similarities and differences between religions, and understand the concept of representation. The aim is that students may learn to understand that the importance of choosing content is also about choosing what is allowed to represent a certain content. Likewise, for example, the ability to reflect and to be able to contextualize one's knowledge are central in the subject of RE. Thus to be able to compare, to understand the concept of representation and to be able to reflect, are three examples of what I mean by powerful knowledge and they are particularly important in religious studies and in the school subject RE. By developing their knowledge and abilities in RE, students can in a longer perspective, through their religious competence, also contribute as active citizens in an inclusive and socially sustainable society (Liljefors Persson, 2022, 2023a, b; Franck & Liljefors Persson, 2023).

9.6 Existential and Controversial Issues in Relation to Religious Education and Sustainable Development Education

The concept of controversial issues has been used in the social sciences for many years and there are many different ways of defining what is meant by it. A simple definition is suggested to be: "Issues that arouse strong emotions and create tension in society" (Kerr & Huddlestone, 2015, p. 13). In other words controversial issues are issues that generate contradictory explanations and complex solutions. These are questions that tend to divide people into different groups, with no easy solutions or answers. They can be both historical and be current in our time. It can be said that they are always sensitive or difficult to deal with and take a stand on. At the same time, these are issues that are significant for learning about and for understanding what democracy is, and for teachers they are considered very challenging to work with in the classroom. Kerr and Huddlestone (2015) articulate the importance of working with controversial issues by emphasizing that it is important for learning about democratic processes and human rights.

The goal of teaching about controversial issues and conducting good discussions in the classroom is to help students develop and test their opinions, gain an increased understanding of, and think about, new angles and perspectives on various issues.

At the same time, they learn to critically review and discuss in a civilized way with their classmates, even if they disagree with the issue (Biesta et al., 2019; Dinham & Shaw, 2017; Kerr & Huddlestone, 2015; Hartsmar & Liljefors Persson, 2013; Flensner et al., 2021; Liljefors Persson, 2023a, b; Franck & Liljefors Persson, 2023).

In teaching about Sustainability and Religious Education, issues of global ethical values, human rights, countering racism and xenophobia and working for inclusion are examples of controversial issues that are important to work with in the classroom and that can contribute to a sustainable society. As a teacher to open discussions between students and to prepare opportunities for meetings between different religions, cultures, norms and values and allowing the students to test their values in the teaching of existential and controversial issues, is urgent and it can contribute in an excellent way to the students' acquiring tools to become active and engaged individuals in a democratic, inclusive and sustainable society.

Working with existential and controversial issues could be an integrated part in Religious Education in connection with questions of Sustainability education and the Agenda 2030. It is also an important part of Citizenship Education (Liljefors Persson, 2009; Cowan & Maitles, 2012; Chistolini et al., 2017). As a teacher, it is important to note conflict-laden opinions, but it is at the same time, somewhat contradictory, important that these are clarified and made visible. Thereby, the students get the opportunity to learn to discuss, understand and respect the opinions of others. This is practicing what democracy is, which is central to what we might call citizenship education, but it is just as well central in Religious Education (Franck & Liljefors Persson, 2023; Liljefors Persson, 2009).

9.7 Concluding Remarks

In this chapter knowledge of Indigenous religions and spirituality, exemplified by the case of Maya cosmology and ritual practices, has been in focus with the aim to deepen the understanding of processes of change and permanence and of sustainability in Indigenous religions. The objective has been to emphasise the importance of integrating questions of sustainability and ecology in RE classrooms at various levels and to encourage future teachers in RE to develop their knowledge on the relations between religion(s) and sustainability.

We have also seen in this chapter that in the long-term Maya religiosity is closely linked to ecology and agricultural life as well as to a holistic perspective on body and health. In the ritual practice of today's live religion, there are thus traces of both duration and change. We can find strong continuities and thus we might emphasis that there is a strong sense of resilience and sustainability in the religiosity and in spiritual landscape among the Maya people over time (Hartsmar & Liljefors Persson, 2013; Liljefors Persson, 2023a, b).

Various Maya peoples have offered fierce resistance to colonialism and we must not limit our view of the Maya to objects that can be easily incorporated into the Western historiography. The Maya have their own historiography, included in the

Chilam Balam Books. Maya culture did not disappear when Europeans arrived in the Yucatan, but it did change. Maya religion and spirituality is dynamic and an ongoing process - where tradition and renewal are constantly reshaping the religious discourse.

In the subject of religious studies, students have the opportunity to compare and reflect on issues of basic values, ethics and morality, as well as a chance to discuss in detail what democracy, solidarity, equality and inclusion can mean. Teachers can, in a didactically conscious way, through existential and controversial issues, connect the teaching about a sustainable society and humanistic values with the general value-based work in the school. If the teaching gives space for existential questions and puts them in relation to human rights and the Convention on the Rights of the Child, the content connects well with the school's value-based work, as well as with humanistic values and what we can call citizenship education.

It has also been suggested here that teaching that is research-based, grounded in powerful knowledge in RE, such as knowledge about indigenous religions, and linked to the concept of a sustainable society, may equip students for active citizenship and gives them the civic education needed to be able to participate in a democratic society. The education that is sought here, may be about embracing global humanistic values to a greater degree and to educate for a sense of justice, generosity, tolerance and global responsibility. In addition, finally, these are questions that may develop students' competence in future existential global issues, such as those we find in Agenda 2030, and thus strengthen their ability to act in a society on the way to greater sustainability.

References

Alberts, W. (2010). The academic study of religions and integrative religious education in Europe. *British Journal of Religious Education, 32*(3), 275–290. https://doi.org/10.1080/01416200.2010.498621

Andersson, D., & Liljefors Persson, B. (2024). Research on indigenous religions. In: G. Larsson & H. Bogdan (Eds.), *The study of religion in Sweden* (pp 1–10). Bloomsbury.

Arzápalo, R. (1987). *El ritual de los Bacabes*. Universidad Nacional.

Barrera-Bassols, N., & Toledo, V. N. (2005). Ethnoecology of the Yucatec Maya: Symbolism, knowledge and management of natural resources. *Journal of Latin American Geography, 4*(12), 9–41.

Biesta, G., Alridge, D., Hannam, P., & Whittle, S. (Eds.). (2019). *Religious literacy: A way forward for religious education?* Brunel University & Hampshire Inspection and Advisory Service.

Bricker, V. (1981). *The Indian Christ, the Indian king: The historical substrate of Maya myth and ritual*. University of Texas Press.

Bricker, V. R., & Miram, H. M. (2002). *An encounter of two worlds: The book of Chilam Balam of Kaua*. Tulane University Press.

Castaneda, Q. E. (2021). "Spiritually seeking the Maya". Toward a decolonial approach to New Age Maya spiritualitites. *Ciencias Sociales y Religión*, 1–37.

Chistolini, S., Holligan, C., Maitles, H., Liduma, A., Liljefors Persson, B., & Papadiamantaki, Y. (2017). *Guidelines for citizenship education in teacher education—Linking research and practice, Children's identity and citizenship in Europe CiCe Jean Monnet network*. London Metropolitan University.

Christenson, A. (2007). *Popol Vuh: Sacred book of the quiche Maya people*. University of Oklahoma Press.

Cotter, C. R., & Robertson, D. G. (2016). Introduction: The world religions paradigm in contemporary religious studies. In C.-R. Cotter & D. G. Robertson (Eds.), *After world religions. Reconstructing religious studies* (pp. xii–xvii). Routledge.

Cowan, P., & Maitles, H. (2012). *Teaching controversial issues in the classroom. Key issues and debates*. Continuum Publishing Corporation.

Craine, E. R., & Reindorp, R. C. (1979). *The codex Perez and the book of Chilam Balam of Mani*. University of Oklahoma Press.

Cusack, C. (2016). Archaeology and the world religions paradigm: The European Neolithic, religion and cultural imperialism. In C.-R. Cotter & D. G. Robertson (Eds.), *After world religions. Reconstructing religious studies* (pp. 153–167). Routledge.

Dinham, A., & Shaw, M. (2017). Religious literacy through religious education: The future of teaching and learning about religion and belief. *Religions, 8*, 119–132. https://doi.org/10.3390/rel18070119

Edmonson, M. S. (1982). *The ancient future of the Itzá: The book of Chilam Balam of Tizimin*. University of Texas Press.

Flensner, K., Larsson, G., & Säljö, R. (Eds.). (2021). *Känsliga frågor, nödvändiga samtal*. Att lära om och av konflikter. Lund.

Franck, O., & Liljefors Persson, B. (2023). Democratic and inclusive religious education in the secular state – The case of Sweden. In F.-V. Anthony & H.-G. Ziebertz (Eds.), *Human rights and the state-religion separation*. Springer.

Gubler, R. (2005). Continuity vs. change in traditional Yucatec curing practices: A tradition in crisis. *Acta Americana, Revista de la Sociedad Sueca de Americanistas, 13*(1–2), 34–54.

Gubler, R. (2017). *Ritos agricolas y ceremonias curativas en Yucatan*. Universidad Nacional Autonoma de Mexico; Gobierno del Estado de Yucatan.

Gubler, R., & Bolles, D. (2000). *The Book of Chilam Balam of Na: Facsimile, translation, and edited text*. Lancaster.

Hall, S. (1992). The West and the rest – Discourse and power. In S. Hall & B. Gieben (Eds.), *Formations of modernity* (pp. 275–332). Polity Press.

Hanks, W. (1986). Authenticity and ambivalence in the texts: A colonial Maya case. *American Ethnologist, 3*(4), 721–744.

Hanks, W. (1990). *Referential practice: Language and lived space among the Maya*. University of Chicago Press.

Hartsmar, N., & Liljefors Persson, B. (Eds.). (2013). *Medborgerlig bildning – demokrati och inkludering för ett hållbart samhälle*. Studentlitteratur.

Heimbrock, H.-G., Scheilke, C. T., & Schreiner, P. (Eds.). (2001). *Towards religious competence: Diversity as a challenge for education in Europe*. Lit verlag.

Hirose Lopez, J. (2015). *Suhuy Maak. Las concepciones sobre el cuerpo y la persona entre los Mayas de la region de los Chenes, Campeche*. Secretario de Cultura de Gobierno del Estado de Campeche.

Houston, S. D., & Inomata, T. (2009). *The classic Maya*. Cambridge University Press.

Jackson, R. (2004). *Rethinking religious education and plurality: Issues in diversity and pedagogy*. Routledge.

Jackson, R. (2008). Teaching about religions in the public sphere: European policy initiatives and the interpretive approach. *Numen, 55*(2008), 151–182.

Jackson, R. (2019). Human rights: A core element or big idea for RE? *Nordidactica – Journal of Humanities and Social Science Education, 4*, 109–132.

Jones, G. (1989). Rebellious prophets. In F. William (Ed.), *Word and image in Maya culture* (pp. 179–193). University of Utah Press.

Kerr, D., & Huddlestone, T. (Eds.). (2015). *Living with controversy – Teaching controversial issues through education for democratic citizenship and human rights (EDC/HRE)*. European Council.

Liljefors Persson, B. (2000). *The legacy of the jaguar prophet. An exploration of Yucatec Maya religion and historiography*. Dissertation. Almqvist & Wiksell.

Liljefors Persson, B. (2006). Blowing in the wind: Divination and curing rituals among the yucatec maya based on the books of chilam balam. *Acta Americana:Journal of the Swedish Americanist Society: Revista de la Sociedad Sueca de Americanistas, Continuities and Changes in Mesoamerican Medicinal Practice, 10*(1–2), 20–33.

Liljefors Persson, B. (2009). Cultural encounters, co-existence and identity – Teaching religious education for a global and sustainable society. In P. Cunningham (Ed.), *Human rights and citizenship education* (pp. 223–228). CiCe.

Liljefors Persson, B. (2011a). "Ualihi Imix Che tu Chumuc": Cosmology, ritual and the power of place in Yucatec Maya (con-)texts. In C. Isendahl & B. Liljefors Persson (Eds.), *Ecology, power, and religion in Maya landscapes* (pp. 145–158). Verlag Anton Saurwein.

Liljefors Persson, B. (2011b). "Zu uinal, zihci kin u kaba, zihci can y luum": Memory, place and ritual in Yucatec Maya (con-)texts. *Acta Americana – Journal of the Swedish Americanist Society, 18*(1–2), 199–213.

Liljefors Persson, B. (2014). U profesia Chilam Balam t ix kayom Cabal Chen Mani: Maya prophecies and cycles of ritual. *Axis Mundi, 9*(1), 78–90.

Liljefors Persson, B. (2017). *"Heix ohemail talci ca chhibal con Maya uinice'*– andlighet, natur, rituella praktiker och förändringsprocesser i Yukatekisk mayareligion". In E. Hall & B. Liljefors Persson (Eds.), *Ursprungsfolkens religioner–perspektiv på kontinuitet och förändring, Föreningen lärare i religionskunskap, FLR's årsbok 2017,* Prinfo Grafiskt tryck, Malmö 2018, 49, pp. 62–80.

Liljefors Persson, B. (2018a). Ursprungsfolkens religioner – perspektiv på kontinuitet och förändring. In E. Hall & B. Liljefors Persson (Eds.), *FLR's årsbok 2017, Årgång 49*. Prinfo Grafiskt tryck.

Liljefors Persson, B. (2018b). Fire symbolism in ritual practices and in ritual speech: Continuity and change in Yucatec Maya (con)texts. In H. Kettunen, V. A. Vázquez López, F. Kupprat, C. Vidal Lorenzo, G. Muñoz Cosme, & M. J. Iglesias Ponce de Leon (Eds.), *Tiempo detenido, tiempo suficiente: Ensayos y narraciones mesoamericanistas en homenaje a Alfonso Lacadena Garcia-Gallo* (pp. 975–990). Wayeb.

Liljefors Persson, B. (2020). (S-)existential questions among students – Sexuality and relations education as part of controversial issues with importance for citizenship education. In B. Krzywosz-Rynkiewicz & V. Zorbas (Eds.), *Citizenship at a crossroads: Rights, identity and education* (pp. 369–381). Charles University; Children's Identity & Citizenship European Association.

Liljefors Persson, B. (2022). "Ämneslitteracitet i SO-undervisning", I Jakobsson, Anders, Nygård Larsson, Pia och Bergman, Lotta (red.). *Ämneslitteracitet och inkluderande undervisning,* Lund: Studentlitteratur, pp. 61–92.

Liljefors Persson, B. (2023a). Undervisning om religion för ett demokratiskt och hållbart samhälle: om ämneslitteracitet och existentiella frågor. In D. Augustsson, C. Carlström, E. Hall, & B. Liljefors Persson (Eds.), *Religion och samhällsförändring – Aktuella perspektiv på religionsvetenskaplig forskning* (pp. 125–146). Stockholm.

Liljefors Persson, B. (2023b). Kosmologi och ekologi i mayas andliga landskap: myter, ritualer och spår av förändring. In D. Augustsson, C. Carlström, E. Hall, & B. Liljefors Persson (Eds.), *Religion och samhällsförändring – Aktuella perspektiv på religionsvetenskaplig forskning* (pp. 165–185). Stockholm.

Love, B. (1986). *Yucatec Maya ritual: A diachronic perspective*. Dissertation. University of California.

Mathews, J., & Garber, J. (2004). Models of cosmic order. Physical expressions of sacred space among the ancient Maya. *Ancient Mesoamerica, 15*(2004), 15–49.

Meyer, J. H. F., & Land, R. (2005). Threshold concepts and troublesome knowledge (2): Epistemological considerations and a conceptual framework for teaching and learning. *Higher Education, 49*, 373–388.

Mulk, I.–M., & Bayliss-Smith, T. (2007). Liminality, art and the sacred Land. *Journal of Northern studies, 1–2*(2007), 95–122.
Redfield, R., & Villa Rojas, A. (1962). *Chan Kom – A Maya village*. The Chicago University Press.
Roys, R. L. (1965). *Ritual of the Bacabs*. University of Oklahoma Press.
Roys, R. L. (1967). *The book of Chilam Balam of Chumayel*. University of Oklahoma Press.
Schele, L., & Mathews, P. (1998). *The code of kings: The language of seven sacred Maya temples and tombs*. Scribner.
Sharer, R. (2009). *Daily life in Maya civilization* (2nd ed.). Greenwood Press.
Shanahan, T., & Shanahan, C. (2012). What is discplinary literacy and why does it matter? *Topics in Language Disorders, 32*(1).
Suarez Castro, M. (2017). El chilam balam de tekax. In *Análisis etnohistórico. Instituto Nacional de Antropología e Historia. Collección Etnohistoria*. Serie Logos.
Taube, K. (2004). Flower mountain: Concepts of life, beauty, and paradise among the classic Maya. *RES Anthropology and Aesthetics, 45*, 69–98.
Taylor, B. (2011). *Dark green religion. Nature spirituality and the planetary future*. University of California Press.
Tozzer, A. M. (Ed.). (1941). *Landa's Relacion de las cosas de Yucatan: A translation* (Papers of the Peabody Museum of American Archaeology and Ethnology 18). Harvard University.
UNEP. (2016). *Environment, religion and culture in the context of the 2030 agenda for the sustainable development*. United Nations Environment Programme, Nairobi.
Williams, K., Hinge, H., & Liljefors Persson, B. (2008). *Religion and citizenship education in Europe*. Children's Identity & Citizenship European Association.
Williams-Beck, L., Liljefors Persson, B., & Anaya Hernández, A. (2012). Back to the future for predicting the past: Cuchcabal – Batabil – Cuchteel and May ritual political structures across archaeological landscapes in ethnohistoric texts, and through cosmological time. *Contributions in New World Archaeology, 4*, 251–278.
World Commission on Environment and Development (WCED). (1987). *Brundtlandrapporten; Report of the World Commission on environment and development: Our common future*. Hämtad 2022-11-01 från: http://www.un-documents.net/our-common-future.pdf
Young, M. (2008). *Bring knowledge back in: From social constructivism to social realism in the sociology of knowledge*. Routledge.
Young, M. (2014). Powerful knowledge as a curriculum principle. In M. Young & D. Lambert (Eds.), *Knowledge and the future: Curriculum and social justice* (pp. 65–88). Bloomsbury.

Bodil Liljefors Persson Professor in History of religions, Malmö University, has for many years conducted research on the religion and history of the Maya people, and regularly does field trips to the Yucatan Peninsula in Mexico. Her research also focuses on religious didactics, sexuality and relations education and citizenship education.

Chapter 10
Religious Education for Sustainable Development in Schools: An Overview of the Topic and Its Didactical Impulses in German-Speaking Countries

Helena Stockinger and Bernd Ziegler

10.1 Introduction

The relatively young concept of Education for Sustainable Development (ESD) has established itself in educational science and educational practice in the last 20 years (cf. Rieckmann, 2021, p. 5). At the beginning of the twenty-first century only a few studies on religious education in German-speaking countries addressed the topic of sustainability (cf. Birkel, 2002; Hisch, 2005; Reis, 2004). In recent years the opportunities and challenges of religious Education for Sustainable Development (rESD) have been increasingly discussed (cf. among others Altmeyer, 2021; Bederna, 2019; Gärtner, 2020; Gärtner et al., 2021; Lehner-Hartmann et al., 2020).

Thus, the discourse on religious education reflects a current trend in society. Various empirical studies in social sciences show that young people in Germany perceive the sustainable transformation of society as a central challenge of modern times. According to a study by the German Federal Environment Agency, environmental and climate protection is a very important or important issue for at least 85 percent of young people (cf. BMUV, 2022, p. 16). The Shell Youth Study from 2019 comes to similar conclusions, with three-quarters of young people in Germany fearing that environmental pollution will have a negative impact on their lives in the future (cf. Schneekloth & Albert, 2019, p. 56). In some cases, they express these fears publicly. The courses of action range from supporting certain political parties in elections to online actions for more sustainable policies to participating in demonstrations of movements such as Fridays for Future (cf. BMUV, 2022, p. 23). Youth opposition to exploitative lifestyles and environmental policies has also

H. Stockinger (✉) · B. Ziegler
Department of Catechetics and Religious Education, Catholic University, Linz, Austria
e-mail: h.stockinger@ku-linz.at

S. Windsor, O. Franck (eds.), *Intersections of Religion, Education, and a Sustainable World*, https://doi.org/10.1007/978-3-031-81809-7_10

influenced current concepts of religious ESD (rESD). They consider the protest of young people as a "sign of formation" (Bederna, 2019, p. 11), which they want to take up productively (All translations from German were done by the authors). Religious education is to become "a sphere in which visions for a good future can be critically designed and the necessary competencies can be acquired to participate in the realization of these visions" (Gärtner, 2020, p. 10).

In the following, the debate on rESD in German-speaking countries will be presented in overview and asks for didactical impulses for teaching at school. Chapter 1 identifies some characteristics of rESD and outlines the current discourse's first controversial issues. Subsequently, concrete learning pathways are named. The conclusion lists some desiderata for the German-language discourse on rESD.

10.2 The Religious Dimension of ESD

Religious ESD is located within the larger framework of Education for Sustainable Development. In the context of school, ESD is mostly conceived as a cross-curricular approach. The concept of Design Competence, which is widespread in German-speaking countries, intends to enable students to gain knowledge and act on sustainability issues in an interdisciplinary way (cf. de Haan, 2008, p. 33). After all, the complex socio-ecological or economic problems of the present can only be adequately grasped through the interaction of numerous disciplines. A similar argumentation is followed by the well-known 'Orientation Framework for Global Development Education in the context of Education for Sustainable Development'. This concept acknowledges the intrinsic value of each subject perspective, but at the same time emphasizes the need for an integrative picture through "interdisciplinary or cross-disciplinary cooperation" (Appelt & Siege, 2016, p. 40). Religious ESD aims to contribute to this big picture. The various models of rESD illustrate in different ways how "religious approaches represent an important dimension of the interdisciplinary ESD project" (Bederna, 2019, p. 210). In doing so, they form their own foci, and these can be condensed into ideal-typical characteristics. In the following, three characteristics are outlined. This does explicitly not exclude the possibility of other forms of presentation or setting priorities.

10.2.1 Normative Orientation

The first characteristic of rESD is its normative orientation. Katrin Bederna's (2019) approach of rESD as messianic transformation education understands sustainability not as a contingent learning object like a lot of ESD approaches in education science, but as a universal moral principle in the sense of the Categorical Imperative (cf. Altmeyer, 2021, p. 59). As such, this concept is strongly grounded in Kantian ethics: Every person is "morally obliged to shape action, life form, society, and economy in such a way that what is thereby done to 'nature' could likewise be done by everyone else everywhere and always" (Bederna, 2019, p. 103). The moral imperative restricts the freedom of action of some people but guarantees a lasting freedom of all those living at present and in the future (Bederna, 2019, p. 185). From a theological perspective, parallels between the sustainability idea and Thomas Pröpper's (2001) concept of freedom emerge. Freedom means to open oneself towards the other and to acknowledge his/her beingness unconditionally (cf. Bederna, 2019, p. 190). Religious ESD aims at the students taking up this moral standpoint – even if the growth and consumption orientation of contemporary society suggests a different decision (cf. Bederna, 2019, p. 233).

In this respect, rESD is a normatively oriented transformation education that "wants to keep alive the question of the necessity of change and make change possible" (Bederna, 2019, p. 236). However, Bederna (2019) emphasizes that rESD does not overwhelm or indoctrinate students. Religious ESD merely seeks to "open up certain attitudes" so that the arguments for sustainability "speak for themselves in their persuasiveness" (Bederna, 2019, p. 233). In any case, manipulative strategies of an unfree transfer of values should be avoided. Transparent communication with clear differentiation between factual information and the views of the teacher, religious community or the like forms the basis for preserving the students' freedom of opinion and judgment (cf. Schimmel & Krahn, 2021, p. 88f.; Gärtner, 2020, p. 115).

Claudia Gärtner's (2020, p. 113) concept of political rESD is also "decidedly normative and partisan". Religious education does not aim at a value-neutral comparison of different sustainability strategies but brings them – spoken from a Christian point of view – in connection with the biblical idea of creation, the messianic message of the Kingdom of God and the alterity of God. Thus, rESD opts for the freedom of all (future) living humans on the planet and contributes to the emancipation of students against social constraints such as strict consumer orientation (cf. Gärtner, 2020, p. 120). Nevertheless, religious education should not just convey supposedly safe solutions to sustainability problems. Such an approach would ignore the fact that "God as the very Other always also eludes human comprehension" (Gärtner, 2020, p. 84). It is rather a matter of uncovering visions, utopias and models of sustainable life from the bible and Christian tradition and offering these to young people as a foundation of meaning for critical appropriation (cf. Gärtner, 2020, p. 110f.). The students are free to decide to what extent they accept this offer.

The normative orientation of rESD is widely accepted in the research discourse on religious education. Simone Birkel's (2002, p. 77) basic idea of a church-based ecological education already argues for "opening up Christian patterns of world interpretation, values and horizons of experience [...] with regard to their ecological relevance and revealing and presenting them as a Christian offer of meaning." Current concepts of rESD continue this normative orientation. This may also be rooted in the concept of (religious) education itself. After all, education aims at a culture of humanity and "sustainability today is part of such a culture" (Vogt, 2021,

p. 702). However, the concrete handling of the normative sustainability ideal is quite debated. Stefan Altmeyer (2021, p. 70), for example, suggests against the background of the heterogeneous perspectives of the students on the ecological crisis to shift the focus away from the introduction of a new normativity towards the promotion of a Translation competence. Especially from a global point of view, it is central to be able to communicate with people of other opinions in the ideologically plural globalized society and to cooperate in overcoming world challenges (cf. Simojoki, 2012, p. 299; Gaus, 2018, p. 344). It is also important to include normative perspectives on sustainability beyond one's own socio-cultural sphere in the educational process – and not to turn subalterns into passive or dependent victims of the ecological crisis (cf. Stockinger & Ziegler, 2023).

Overall, a tension in dealing with normativity is inherent in rESD. On the one hand, the ecological crisis seems to make orientation towards a concept of freedom necessary, which limits the freedom of action of the individual in favor of the freedom of all (cf. Bederna & Gärtner, 2022, p. 23). On the other hand, values such as sustainability, when played into religious education in a normative way, contain the danger of an unfree transfer of values to the students (cf. Grümme, 2021, p. 347). Constructively dealing with this tension remains a central challenge for the theory and practice of rESD.

10.2.2 References to Religious Traditions and Theological Argumentation

A second characteristic has already been mentioned in the context of normative orientation. Religious ESD is defined by explicit references to religious traditions and theological argumentation. Especially in the school context, this results in the challenge to bring a "genuinely theological or religious perspective controversially into the discourses on sustainability" (Gärtner, 2020, p. 39).

Bederna's concept of messianic transformation education emphasizes, from a Catholic perspective, that the emotional and communal quality of a relationship with God can positively influence sustainable action. Christianity offers a potential for hope and utopia to help shape the future (cf. Bederna, 2019, p. 111). First, religious education brings students into contact with transformation-relevant religious knowledge (cf. Bederna, 2019, p. 239). This knowledge can also be partly scientific, economic or similar. However, it is always integrated into "the 'dangerous memory' (Johann Baptist Metz) of the life, death and resurrection of Jesus, which transformed half the world, and from which the anticipation of a transformed future grew and can continue to grow" (Bederna, 2019, p. 242). This includes, among other things, knowledge of biblical creation texts, a sustainability-related view of alternative ways of life from Christian motivation such as the poverty ideal of the twelfth/thirteenth century, reflections on the relationship between sustainability and freedom as well as concrete sustainability efforts of different religious communities (cf. Bederna, 2019, p. 241). Secondly, religious education can promote transformation-relevant religious competencies. Particularly important here is a perceptual competence (cf. Bederna, 2019, p. 244). Ethical-moral competencies are also central, such as the change of perspective to develop a moral point of view, or participation competencies to contribute to the option for the poor and creation (cf. Bederna, 2019, p. 255). Thirdly, religious education provides a space for the formation of creation spirituality (cf. Bederna, 2019, p. 246; Hisch, 2020, p. 95). The performance of certain creation texts could motivate students to want to restore God's good order and thus to act sustainably (cf. Bederna, 2019, p. 250). Perhaps religious education even has some potential to play a role in overcoming the mind-behavior gap between well-grounded knowledge about the ecological crisis and non-sustainable behavior (cf. Bederna & Gärtner, 2022, p. 21; Altmeyer & Stockinger, 2019).

Birkel's (2019, p. 180) reflections on religious education and sustainability also emphasize the motivating character of religion. She proposes to establish the "principle of gratuity" – i.e., giving a gift free of charge – as a religious contribution to ESD: Christians have received the good of creation for free and are therefore called to pass on the good gratuitously. Religious education can promote the "lust for a good life" (Birkel, 2019, p. 179). Important elements in this regard are the perception of and wonder about nature, one's own creative activity in everyday life, cognitive reflection on the background of the ecological crisis, and communication and reinforcement about the possibility of a different world in the church community (cf. Birkel, 2016, pp. 10–13). Such education does not only take place in a specific lesson but requires a holistic restructuring of school as a place of learning in the sense of the Whole School Approach (cf. Birkel, 2021, p. 121; differently accented in Bederna, 2019, p. 252; see also chapter 3).

Gärtner (2020, p. 135) highlights another religious reference, which is also briefly mentioned by Bederna: Political rESD has an "ideology-critical potential". Christian beliefs, for example, can make people aware of the eschatological limitations of human action. From this perspective, human agency that presents itself as a safe and alternative-less way to a better world is to be considered problematic. In religious education, students can be encouraged to critically examine utopias within the sustainability discourse, such as the supposed cure-all 'Green Economy', for their ideological potentials (cf. Gärtner, 2020, p. 115). Finally, ideology critique should be self-reflexive: To what extent has Christianity promoted an exploitative anthropocentrism? And in what ways does religious education also contribute to current sustainability problems (cf. Gärtner, 2020, p. 123)?

The ideology-critical moment of rESD already indicates that sustainability issues are always structural issues as well (cf. Bederna, 2019, p. 264). In this respect, it is the purpose of religious education to grasp sustainability also from a political perspective.

10.2.3 *Political Dimension*

Religious ESD is politically dimensioned. Otherwise, there is a danger that sustainability issues will be overly individualized (cf. Gärtner, 2020, p. 27). This would make students uncritical of the concrete political implementation of sustainability. The conflicts between ecology, economy and social issues as well as between the Global South and the North, which come to light in the political debate, should not remain underexposed. On the contrary, they should be explicitly addressed and discussed in the classroom (cf. Gärtner, 2020, p. 27; Bederna, 2019, pp. 128–131; Vogt, 2021, p. 680). Nevertheless, quite a few teaching materials for religious education tend to show the opposite tendency. Here creation theology is often shortened in individual ethical terms to the call to collect garbage (cf. Benk, 2018; Bederna, 2019, pp. 160–178; Gärtner, 2020, p. 29).

There are also theological arguments for a political dimension of rESD. From a Christian perspective, rESD refers to the fact that Jesus' message of the Kingdom of God has political connotations. Structural questions are therefore an important topic of religious education (cf. Bederna, 2019, p. 250). In the classroom, sustainability issues can then be grasped, for example, with the concept of structural sin described in the Second Vatican Council: People live in unsustainable social structures that make it very difficult for the individual person to live sustainably (cf. Gärtner, 2020, p. 91). These structures can be analyzed and criticized in religious education, taking into account their complexity. Thus, religious education contributes to the emancipation of students from structural constraints in favor of dealing with nature in a sustainable way (cf. Gärtner, 2020, p. 120f.; Bederna, 2019, p. 257). Religious ESD denies – as it is also demanded in creation theology – that there is no alternative for the current policy of exploiting nature (cf. Benk, 2016, pp. 253–270). Students are motivated to work for a good world related to God (cf. Gärtner, 2020, p. 94; Bederna, 2019, p. 167; Benk, 2016, pp. 253–270).

Overall, the normative orientation, the religious and theological points of reference, as well as the strong political focus are central characteristics of the current concepts of rESD. This is also evident in the concrete didactic ideas for learning and teaching in schools.

10.3 Didactical Concretizations of rESD

In the German-speaking world, there are currently two profiled didactic approaches to rESD in schools. These concepts by Gärtner (2020) and Bederna (2019) develop some didactic contours or principles from their theoretical considerations, which they also transfer into practical impulses for learning and teaching. In addition, other approaches – even if they are not based on an explicit didactic concept for school education – also name different concretizations or methods of how rESD can succeed in the classroom. Some approaches aim directly at changing an individual lifestyle (cf. e.g. Birkel, 2002), others more strongly at working on the structural component of sustainability (cf. e.g. Gärtner, 2020). Most of the didactical impulses focus on the specific contribution of religious education to creating a more sustainable society, whereas a few ideas explicitly address the whole school (cf. Birkel, 2016). From these different concepts, some general didactic guidelines can be identified. Based on that, the heterogeneous methods and didactical concretization of rESD can be systematized.

10.3.1 Perception-Oriented Approaches

Birkel (2016, p. 10) emphasizes that perceiving nature with all senses is fundamental for individuals to form an attention oriented towards sustainability. Bederna (2021, p. 261), in contrast, critically states that experiencing nature does not play a role in the general ESD concept of Design Competence. At the same time, both Bederna and Gärtner emphasize the importance of aesthetic approaches, whereby this is linked to political implications. In her three-step approach, Gärtner names the critical perception of sustainability as the first step, in which the respective perspectives and contexts must be taken into account (cf. Gärtner, 2021, p. 78). Bederna (2019, p. 244) also mentions the importance of training perceptual competence by discovering problems of non-sustainability with students. In addition, performative elements, such as the performance of creation texts, can motivate students to act sustainably (cf. Bederna, 2019, p. 250).

10.3.2 Experiential and Action-Oriented Approaches

Particular emphasis is placed on experiential and action-oriented approaches, which can be found in different forms in all concepts of rESD.

10.3.2.1 ESD Service Learning, Compassion and Diaconal Learning, Civil Society Projects

Bederna (2019, pp. 267–272) emphasizes the importance of ESD service learning. She distinguishes this concept from compassion or diaconal learning, as she does not perceive them as transformative and politically dimensioned like ESD Service Learning. The purpose of ESD service learning is to change structures. ESD Service Learning is not institutionally embedded and is an instructionally open way to deal with sustainability issues. In its ideal form, ESD Service Learning follows the critique phase of a future workshop. Unlike other forms of service learning, ESD service learning should begin with the problem that is being attempted to be addressed. The planning phase, which is essential in this context, is largely left to the students, and planned projects should allow experiences of self-efficacy. The project should have a clear goal and conclusion and be time-limited. The effects of the project should be evaluated afterwards. Reflection, which is a central characteristic of service learning, should be both theological and political and must not individualize. Success is demonstrated less by presentable results than by how resistance and crises are reflected upon and what creative responses are found. Recognition through awards for creative or effective sustainability projects is also part of service learning (cf. Bederna, 2019, p. 270). As concrete examples that have arisen in the context of service learning, Bederna cites that students have cleaned a lakeshore from plastic and have put up a sign for plastic avoidance. Other projects include supporting a Critical Mass, providing free rental bikes, setting up a ride-sharing app, creating a climate repentance service, designing a memorial wall for climate change victims, etc. Even if religious references are not explicitly evident in some projects, the projects can be interpreted "as an experience of self-transcendence and commitment to the kingdom of God" (Bederna, 2019, p. 271).

Gärtner (2020, S. 128f.) mentions the religious learning pathway of 'Compassion'. Compassion projects, which are internships in social institutions, could also promote the perception and judging of sustainability issues if they are carried out in cooperation with nature conservation organizations or fair trade companies. After the project, the students' experiences are reflected in detail. However, the concept of compassion must also be critically questioned: How are students supposed to develop compassion for people from the Global South or, for example, war refugees, if they themselves cannot have any corresponding experiences?

Schimmel and Krahn (2021, p. 92) show that even small steps can be a motivating start. Specific individual areas can be selected for a personal experiment of changed behavior. Besides the identification of the ecological footprint, civic-political activities for the promotion of a sustainable society can be targeted. In contrast, Birkel (2016, p. 11) aims more at the individual actions of the people. Thus, she mentions different activities that young people can do, such as producing homemade things and experiencing themselves independently and creatively.

10.3.2.2 Experience of Community

Birkel (2019, p. 12) emphasizes the importance of community that shares the vision of a different world. Positive group experiences in the implementation of commitment to sustainability are likely to have a strong effect on the development of attitudes (cf. Schimmel & Krahn, 2021, p. 92), as this also addresses the affective dimension of attitudes. The importance of community is also emphasized by Gärtner (2020, p. 126). According to this, community-based forms of learning and the discussion of the values and norms of a religious community can have a beneficial effect on students' sustainable action.

10.3.2.3 Exploring Places and Encounter-Learning

Gärtner (2021, p. 80) particularly emphasizes the importance of encouraging examples of Christian action. In a monastery garden, sustainable living in religious tradition can be learned both physical-aesthetically and spiritually-liturgically (cf. Gärtner, 2020, p. 140f.). Places like these are in tension with society both structurally-institutionally and biographically-individually, which can result in dilemma situations.

As another possibility of experiential learning, Schimmel and Krahn (2021, p. 92) address encounter learning. Thus, authentic encounters with peers from the Global South can contribute to questioning one's own understanding of "normal" (Schimmel & Krahn, 2021, p. 92). The change of perspective made possible by this can sensitize empathy for suffering and underlying injustices, from which responsibility in solidarity can also arise (cf. Gaus, 2014, p. 38). References between Global Education and ESD become evident here. Gärtner (2021, p. 80) also emphasizes the possibility of establishing contact with groups such as Christians for Future or Christian activists. On the one hand, this can serve to encourage sustainable action, but on the other hand, it can also lead to a critical reflection of the experienced sustainable practice.

10.3.2.4 Connection to Life-World Contexts and Crises

Gärtner (2021, p. 78) emphasizes that motivating rESD learning settings often link to students' lifeworld contexts. For example, the Lenten season, during which abstinence is practiced by many people for various motives, could be a context to reflect on sustainable lifestyles in its individual as well as its social, political, and spiritual dimensions (cf. Gärtner, 2020, p. 142). In addition, experiences of crises provide learning opportunities for adolescents to critically reflect on their own way of life and to consider what a good, shared life for all might look like (cf. Gärtner, 2020, p. 158). This requires the verbalization and reflection of experiences. In a second step, adolescents can exchange these experiences and thus recognize that they are not alone in their experiences. At the same time, the exchange of experiences also reveals differences that can be further processed in a heterogeneity-sensitive manner. If lifeworld- and context-oriented dilemmas are addressed, problems can be recognized and discussed as well as approaches to solutions (cf. Gärtner, 2020, p. 154f) In addition, the tension between individual and structural entanglements becomes apparent in many everyday situations, which can be addressed in terms of religious education.

10.3.3 Communication About Different Topics

In addition, various forms of how students get into conversation with each other about sustainability are discussed. The goal here is often to change perspectives and thereby understand other positions.

10.3.3.1 Starting from Theological Considerations

Theological reflections can open horizons for adolescents or provide interpretations. According to Gärtner (2021, p. 76) tensions, but also the feeling of futility can be described particularly well with anamnetic and eschatological figures of thought. In Christianity, the memory of the saving deeds of Jesus Christ offers hope for change. At the same time, the message of the Kingdom of God makes clear that human practice is always an unfinished practice and can open a hermeneutical and utopian horizon. Through concrete parables such as the growth or contrast parable (Mk 4:30–32), young people from grade 9 onwards can develop aspects of the message of the Kingdom of God. The discussion about the kingdom of God is particularly worthwhile because it is characterized by tensions and offers visions for the future. According to Gärtner, this discussion of theological content is secondary to the critical perception of sustainability and prior to the discussion of encouraging examples of Christian action.

Schimmel and Krahn (2021, p. 93) also emphasize that it is necessary to interpret central biblical topoi such as creation theology as well as Jesus' message of the kingdom of God as a better world in the context of climate issues. The importance of dealing with creation theology and faith in creation is also emphasized by Ziermann (2021, p. 115). Likewise, a study of ecclesiastical teaching letters of the Protestant and Catholic Churches, including the encyclical "Laudato Si" (2015), can take place. For students to understand a theological (world) interpretation, aspects of theological anthropology such as responsibility before God or Dietrich Bonhoeffer's ethics of responsibility can be addressed (cf. Schimmel & Krahn, 2021, p. 93). Lena Tacke (2020) explains learning sequences in which students, based on the discussion of prophets, ask the question to what extent Greta Thunberg can be described as one in terms of form and content. Subsequently, a connection can be made between Greta Thunberg's call for the integrity of creation and this very commandment of God. Accordingly, Christian responsibility in the face of climate change can be reflected upon (cf. Tacke, 2020, p. 124).

Bederna (2021) sees the learning pathway of theologizing with students as an important contribution to rESD. Accordingly, the communal-sacramental-aesthetic-temporal dimensions grouped around the space of theologizing can help to promote a new relationship to nature and the future and to overcome the gap between knowledge and action (cf. Bederna, 2021, pp. 64 f.). In religious education, students' experiences of the ecological crisis can be addressed and reflected in all dimensions. Such theologizing ideally leads to a knowledge of different reasons for sustainability and thus to better intuitions as well as to a different normality (cf. Bederna, 2021, p. 69).

10.3.3.2 Diverse Discourse Stimuli

Besides theologizing, conversations and dilemma discussions can be further discursive methods (cf. Bederna, 2019, p. 263). Considering the ethical dimension of Christianity, Gärtner (2020) emphasizes the importance of value development and communication. In this context she cites dilemma situations as a popular methodological learning setting (cf. Gärtner, 2020, p. 129f). In working on dilemma situations, it becomes clear that especially in conflicts of conscience the clarification of the concept of structural sin can be an aid to interpretation (cf. Gärtner, 2020, p. 154) Based on the concept of structural sin, the dilemma can be processed as structural and social. The method of framing can also be used to introduce students to different interpretive perspectives (cf. Gärtner, 2020, p. 119). Here, different "reading glasses" are tried out with adolescents, whereby texts, situations or problems are interpreted and discussed from different perspectives in an ideology-critical way. These different frames enable the adolescents to analyze how reality can be (de) constructed. Following the framing method, students can also be creative by expressing fears, worries, or hopes in texts, songs, or pictures and sharing this with others, depending on the learning group.

Discourse-opening topics can also vary. Social concepts such as Buen Vivir (cf. Welthaus Bielefeld, 2012; Rieckmann, 2017), but also socio-analytical approaches such as the theory of the acceleration society and the search for resonance (Hartmut Rosa) or the sufficiency idea in the context of a post-growth economy (Niko Paech) can be dealt with (cf. Schimmel & Krahn, 2021, p. 93) Based on the Latin American concept of Buen Vivir, individuals can formulate conditions of success of a good life in the face of an achievement-oriented society, whereby here environment and human beings are to be considered globally and regionally (cf. Birkel, 2016, p. 13). This idea is also taken up by Gärtner (2020, p. 160).

Schimmel and Krahn (2021, p. 91) also emphasize the importance of responding to young people's values. Thus, after a phase of self-explanation in individual work and small groups, expressed aspects can be addressed in the group and discourse impulses can be offered. Addressed emotions should be taken seriously in this context.

10.3.4 Importance of Content-Related Knowledge

In addition, the importance of the cognitive component is emphasized in some approaches. It takes content knowledge about lifestyles, as well as scientific knowledge, to enable attitude change. According to Birkel (2016, p. 12), for a new lifestyle to be shaped, it requires knowledge of innovative sociopolitical approaches and life models. Schimmel and Krahn (2021, p. 90) also emphasize the need for secure scientific knowledge on climate change as a necessary element for a problem-conscious attitude towards climate change. This also means being able to correctly bring together scientific facts, ethical and religious considerations, etc. in their complexity. To do this, students need the ability of system thinking (cf. Bederna, 2019, p. 261; Holzwieser, 2020, p. 106f.). In religious education, this can be promoted, for example, by the mystery method (cf. Bederna 2019, p. 261). In this method, potentially relevant facts are presented on small cards, which are to be brought into a context to solve a problem. In sustainability issues, the number of small cards will never be complete and the contents are to be requested by the students themselves. Furthermore, students should learn about socio-psychological processes of opinion formation (cf. Schimmel & Krahn, 2021, p. 90). Knowledge about the ideal of poverty or efforts towards sustainability of individual religious communities can also be broadened in religious education (cf. Bederna, 2019, p. 241).

10.3.5 Structural Integration in the Overall School Context

Birkel (2021) emphasizes the importance of the holistic restructuring of the school as a place of learning, which she elaborates with the Whole Institution Approach. Common characteristics of this approach, which may be manifested differently at schools, are sustainability as part of everyday actions, consideration of management and operations, a focus on staff and continuing education, the importance of organizational and leadership culture, internal and external communication, and regional and transregional collaborations (cf. Birkel, 2021, p. 122). In this context, religious education can contribute to the holistic orientation of the school. Here, for example, Birkel addresses rituals and liturgical elements in terms of aesthetic education. In her opinion, biblical stories can also give rise to visions of a just world in the sense of storytelling. In the context of the message of the Kingdom of God, Christians can point to a "creation-responsible, participatory, transparent, and tolerant way of living with all fellow creatures" (Birkel, 2021, p. 124).

The importance of cooperation between different subjects is also mentioned. Based on their qualitative-empirical study on the decision-making behavior of adolescents in ecological questions, Altmeyer and Dreesmann name the importance of subjects such as religion and biology cooperating with each other (cf. Altmeyer & Dreesmann, 2020, p. 301) and name concrete suggestions for interdisciplinary learning (cf. Altmeyer & Dreesmann, 2018).

10.4 Desiderata

The debate on rESD in the German-speaking world reveals the small number of empirical projects that both test the hypotheses raised in theory and develop theory from practice. The projects so far aim at a strongly individual approach to the problem of sustainability. In a quantitative study of 14- to 16-year-old students in 18 secondary schools in central-western Germany, individuals are asked to position themselves on a dilemma using a detailed questionnaire (cf. Altmeyer & Dreesmann, 2018, 2020). The study seeks to elicit the possible role of religious orientations in adolescents' decisions in an ecological dilemma. Another empirical project represents a case study of how adolescents deal with the category of (un)certainty of ecological crisis and societal future (cf. Torrau & Gloe, 2021). Not directly related to religious ESD – also due to its temporal location – is a study by Boris Kalbheim (2000). The empirical study on the motives of church and non-church members to act in an environmentally friendly manner can hardly be used to draw conclusions for the concretization of rESD (cf. also Birkel, 2002, 75). In another project, five interviews were conducted with people who are or were active in the environmental team of their church congregation (cf. Ziermann, 2021). In the interviews, the strong reference to the theology of creation and the belief in creation resonates. Creation is also the subject of Christian Höger's (2020) study, which surveys the attitudes of schoolchildren towards creation, the big bang and evolution in a long-term study. Again, there is no direct reference to rESD.

What becomes clear in the overview of empirical studies on religious education in the context of the discourse on rESD is that there is a considerable lack of empirical studies. There's no data on how rESD can be implemented in schools and what the effects of various learning settings are. Especially in view of the manifold concretisations cited in the theoretical discourse, some of which are consistently derived from the preceding theory or some of which seem to be arbitrarily set, the question arises as to the effectiveness of the concretizations. In this context, it also seems useful to differentiate which competencies people are supposed to acquire through certain learning settings. Connected to this is the question to what extent ESD can be linked to well-known principles of religious didactics such as learning church history, biographical learning or anamnetic learning. The possibility of applying these principles in the discourse on ESD is noted in various places, but not developed in more detail. What is needed here is a differentiated clarification of which principles religious ESD relates to and how, and how didactical principles for teaching religion are to be adapted in relation to rESD.

Three further research desiderata should be emphasized in conclusion: the discourse on rESD has been examined from a Christian perspective; interreligious perspectives have been lacking up to now. Here it is necessary to ask to what extent there is ESD from the perspective of different religious approaches and to bring together different figures of thought. It would be interesting here to work out a concept of rESD across religious boundaries. In addition, there are open topics and problem areas that are partly addressed but not yet developed in detail. For example, thematic areas of the postcolonial discourse, which are already received in the context of Global Education (cf. Simojoki, 2018; Henningsen, 2022), have hardly been unfolded yet in the discourse on rESD in German speaking countries (Stockinger & Ziegler, 2023). Moreover, the discourse on rESD increasingly refers to German-language literature. Here, it would still be fruitful for the discourse if more international findings were included in religious education considerations.

References

Altmeyer, S. (2021). Religious education for ecological sustainability: An initial reality check using the example of everyday decision-making. *Journal of Religious Education, 69*(1), 57–74. https://doi.org/10.1007/s40839-020-00131-5

Altmeyer, S., & Dreesmann, D. (2018). Grenzgänge zwischen Natur und Schöpfung. Grundlagen und Vorschläge für fächerverbindendes Lernen in Biologie- und Religionsunterricht. *Jahrbuch der Religionspädagogik, 34*, 171–183.

Altmeyer, S., & Dreesmann, D. (2020). The importance of religion for the evaluation of everyday ecological decisions by German adolescents. *Worldviews: Global Religions, Culture, and Ecology, 24*(3), 285–307. https://doi.org/10.1163/15685357-20203001

Altmeyer, S., & Stockinger, H. (2019). Ökologisches Lernen IST religiöses Lernen. *Katechetische Blätter, 144*(3), 165–167.

Appelt, D., & Siege, H. (2016). Konzeptionelle Grundlagen des Orientierungsrahmens. In BMZ & KMK (Eds.), *Orientierungsrahmen für den Lernbereich globale Entwicklung im Rahmen einer Bildung für nachhaltige Entwicklung* (2nd ed., pp. 21–54). Cornelsen.

Bederna, K. (2019). *Every day for future: Theologie und religiöse Bildung für nachhaltige Entwicklung*. Grünewald.

Bederna, K. (2021). Didaktik religiöser Bildung für nachhaltige Entwicklung. In U. Kropac & U. Riegel (Eds.), *Handbuch Religionsdidaktik* (pp. 325–331). Kohlhammer.

Bederna, K., & Gärtner, C. (2022). Dramatisch! Irrelevant? Gott suchen, erfahrungsbezogen theologisieren und solidarisch unterbrechen. Fünf Thesen zu religiöser Bildung für nachhaltige Entwicklung. *Loccumer Pelikan, 2*, 18–24.

Benk, A. (2016). *Schöpfung – eine Vision von Gerechtigkeit: Was niemals war, doch möglich ist.* Grünewald.

Benk, A. (2018). Schöpfung: trivialisiert, separiert, historisiert und instrumentalisiert – oder eingebunden in den befreienden Horizont biblischer Hoffnung? Kritische Sichtung unterrichtspraktischer Materialien zur Schöpfungsthematik. In S. Altmeyer, R. Englert, H. Kohler-Spiegel, E. Naurath, B. Schröder, & F. Schweitzer (Eds.), *Jahrbuch der Religionspädagogik: Vol. 34. Schöpfung* (pp. 229–248). Vandenhoeck & Ruprecht.

Birkel, S. (2002). *Zukunft wagen – ökologisch handeln: Grundlagen und Leitbilder kirchlich-ökologischer Bildung im Kontext nachhaltiger Entwicklung*. Lit.

Birkel, S. (2016). Die Sorge für das gemeinsame Haus: Herausforderungen für die Religionspädagogik durch die Mitwelt-Enzyklika Laudato si. *Religionspädagogische Beiträge, 75*, 5–13.

Birkel, S. (2019). Warum wir dran sind: Lernen für ein zukunftsfähiges Leben. *Katechetische Blätter, 144*(3), 177–181.

Birkel, S. (2021). "Generation Greta" – Herausforderungen für Religionsunterricht und Schule im Kontext eines Whole Institution Approach (WIA). *Religionspädagogische Beiträge, 44*(2), 117–126. https://doi.org/10.20377/rpb-143

Bundesministerium für Umwelt, Naturschutz, nukleare Sicherheit und Verbraucherschutz (Ed.). (2022). *Zukunft? Jugend fragen! Umwelt, Klima, Wandel – was junge Menschen erwarten und wie sie sich engagieren*. Bundesministerium für Umwelt, Naturschutz, nukleare Sicherheit und Verbraucherschutz.

de Haan, G. (2008). Gestaltungskompetenz als Kompetenzkonzept. In I. Boormann & G. de Haan (Eds.), *Kompetenzen der Bildung für nachhaltige Entwicklung: Operationalisierung, Messung, Rahmenbedingungen, Befunde* (pp. 23–43). VS Verlag für Sozialwissenschaften.

Gärtner, C. (2020). *Klima, Corona und das Christentum: Religiöse Bildung für nachhaltige Entwicklung in einer verwundeten Welt*. transcript.

Gärtner, C. (2021). Alles vergeblich!? Religionsdidaktische Konkretionen einer politischen religiösen Bildung für nachhaltige Entwicklung. *Religionspädagogische Beiträge, 44*(2), 73–83. https://doi.org/10.20377/rpb-118

Gärtner, C., Knauth, T., & Stockinger, H. (2021). Herausforderungen für (religiöse) Bildung angesichts der ökologischen Krise und einer gänzlich offenen Zukunft der Welt. *Religionspädagogische Beiträge, 44*(2).

Gaus, R. (2014). Gerechtigkeit und Wirtschaft: Option für die Armen im Wirtschaftssystem des globalen Marktes: Entwurf für die Oberstufe. *ReliS*, (1), 42–47.

Gaus, R. (2018). Globales Lernen im Religionsunterricht. In J. Sautermeister & E. Zwick (Eds.), *Religion und Bildung: Antipoden oder Weggefährten?* (pp. 237–247). Schöningh.

Grümme, B. (2021). *Praxeologie. Eine religionspädagogische Selbstaufklärung*. Herder.

Henningsen, J. (2022). *Repräsentationen des Globalen Südens im evangelischen Religionsbuch: Eine Thematische Diskursanalyse vor dem Horizont postkolonialer Theorien*. Schöningh.

Hisch, J. (2005). Das Thema Nachhaltigkeit als gemeinsame religionspädagogische Herausforderung. *Österreichisches Religionspädagogisches Forum, 15*(1), 33–34.

Hisch, J. (2020). Spiritualität und Bildung für Nachhaltige Entwicklung bei PILGRIM. *Österreichisches Religionspädagogisches Forum, 28*(2), 82–99. https://doi.org/10.25364/10.28:2020.2.6

Höger, C. (2020). *Schöpfung, Urknall und Evolution – Einstellungen von Schüler*innen im biographischen Wandel: Ein qualitativ-empirischer Längsschnitt mit dem Ziel religionspädagogischer Pünktlichkeit im Religionsunterricht der Sekundarstufe*. Lit.

Holzwieser, M. (2020). Grüne Pädagogik und konstruktivistische Religionspädagogik: Ein Vergleich zweier konstruktivistischer pädagogischer Konzepte. *Österreichisches Religionspädagogisches Forum, 28*(2), 100–116. https://doi.org/10.25364/10.28:2020.2.7

Kalbheim, B. (2000). *Sinngebung der Natur und ökologisches Handeln: Eine empirisch theologische Untersuchung zu Motiven umweltschützenden Handelns bei Kirchenmitgliedern und NichtkirchenmitgliedernEmpirische Theologie: Vol. 4*. Lit.

Lehner-Hartmann, A., Dangl, O., & Rothgangel, M. (2020). Nachhaltiges Lernen. *Österreichisches Religionspädagogisches Forum, 28*(2).

Pröpper, T. (2001). *Evangelium und freie Vernunft: Konturen einer theologischen Hermeneutik*. Herder.

Reis, O. (2004). *Nachhaltigkeit - Ethik - Theologie: Eine theologische Beobachtung der Nachhaltigkeitsdebatte*. Lit.

Rieckmann, M. (2017). Bildung für nachhaltige Entwicklung in der Großen Transformation: Neue Perspektiven aus den Buen Vivir – Postwachstumsdiskursen. In O. Emde, U. Jakubczyk, B. Kappes, & B. Overwien (Eds.), *Schriftenreihe "Ökologie und Erziehungswissenschaft" der Kommission Bildung für eine nachhaltige Entwicklung der DGfE. Mit Bildung die Welt verändern? Globales Lernen für eine nachhaltige Entwicklung* (pp. 147–159). Barbra Budrich.

Rieckmann, M. (2021). Reflexion einer Bildung für nachhaltige Entwicklung aus bildungstheoretischer Perspektive. *Religionspädagogische Beiträge, 44*(2), 5–16. https://doi.org/10.20377/rpb-153

Schimmel, A., & Krahn, A. (2021). "Klima nervt!": Zum didaktischen Umgang mit Widerständen bei der Thematisierung des Klimawandels im Religionsunterricht. *Religionspädagogische Beiträge, 44*(2), 85–96. https://doi.org/10.20377/rpb-154

Schneekloth, U., & Albert, G. (2019). Jugend und Politik: Demokratieverständnis und politisches Interesse im Spannungsfeld von Vielfalt, Toleranz und Populismus. In M. Albert, K. Hurrelmann,

G. Quenzel, U. Schneekloth, I. Leven, & H. Utzmann (Eds.), *Shell-Jugendstudie. Jugend 2019: Eine Generation meldet sich zu Wort* (pp. 47–101).

Simojoki, H. (2012). *Globalisierte Religion: Ausgangspunkte, Maßstäbe und Perspektiven religiöser Bildung in der Weltgesellschaft. Praktische Theologie in Geschichte und Gegenwart: Vol. 12*. Mohr Siebeck.

Simojoki, H. (2018). Ökumenisches Lernen, Hybridisierung und Postkolonialismus: Versuch einer kritischen Verschränkung. In A. Nehring & S. Wiesgickl (Eds.), *Postkoloniale Theologien: Vol. 2. Perspektiven aus dem deutschsprachigen Raum* (pp. 256–270). Kohlhammer.

Stockinger, H., & Ziegler, B. (2023). Postkoloniale Perspektiven auf religiöse Bildung für nachhaltige Entwicklung. *Österreichisches Religionspädagogisches Forum, 31*(1), 106–126. https://doi.org/10.25364/10.31:2023.1.7

Tacke, L. (2020). „Nur noch kurz die Welt retten…"? Ermutigung und Befähigung zueinem nachhaltigen Lebensstil durch den Religionsunterricht. Eine Unterrichtsidee. *Österreichisches Religionspädagogisches Forum, 28/ 2*(2), S. 117–131. https://doi.org/10.25364/10.28:2020.2.8

Torrau, S., & Gloe, M. (2021). Es sind nun mal leider keine Klimaaktivisten, die das Land führen. *Religionspädagogische Beiträge, 44*(2), 127–139. https://doi.org/10.20377/rpb-149

Vogt, M. (2021). *Christliche Umweltethik: Grundlagen und zentrale Herausforderungen*. Herder.

Welthaus Bielefeld (Ed.). (2012). *"Buen Vivir - Was heißt Gutes Leben?" Bildungsmaterial für die Fächer Praktische Philosophie und Religion (Kl. 9+10)*. Welthaus Bielefeld.

Ziermann, S. (2021). Eine studentische Studie zur kirchlichen Umweltarbeit oder: Schöpfungstheologie im Umweltschutz entdecken. *Religionspädagogische Beiträge, 44*(2), 107–116. https://doi.org/10.20377/rpb-140

Helena Stockinger is Professor of Catechetics and Religious Education at the Catholic University in Linz, Austria. She has held teaching and research positions in different countries, e.g. as professor of Religious Education at the University of Munich and as a visiting scholar at Boston College. She is vice-chair of the German-speaking Religious Education Association and member of the editorial board of the Journal for Religion in Education.

Bernd Ziegler is Assistant Professor at the department of Catechetics and Religious Education at the Catholic University in Linz, Austria. He has held different teaching and research positions, for example as a postdoctoral researcher at the University of Munich and as a visiting scholar at the Pontifical Catholic University of Paraná. He is a member of the working group developing the German "Curriculum Framework: Education for Sustainable Development".

Chapter 11
What Matters to Senior Secondary Students in RE: An Australian Case Study That Compares an Innovative Curriculum with Traditional Approaches

William Sultmann, Janeen Lamb, Peter Ivers, and Mark Craig

11.1 Introduction

Religious Education plays a central role in achieving the church's mission (Congregation for Catholic Education [CCE], 2009). This perspective indicates that Religious Education in Catholic schools is not an optional extra but an integral part of the school curriculum because 'without religious education, pupils would be deprived of an essential element of their formation and personal development, which helps them attain a vital harmony between faith and culture' (CCE, 2009, para. 10). The importance of Religious Education for Catholic schools in Australia is reinforced in the National Catholic Education Commission (NCEC) Framing Paper: Religious Education in Australian Catholic Schools (NCEC, 2018), with the fundamental and challenging goals of developing:

> … students' knowledge and understandings of Christianity in the light of Jesus and the Gospel, and its unfolding story and diversity within contemporary and global society. It expands spiritual awareness and religious identity, fostering their capacity and skills discerning, interpreting, thinking critically, seeking truth and making meaning. It challenges and inspires their service to others and engagement in the Church and the world (NCEC, 2018, p. 7).

The purpose of this chapter is to trace the current status of two approaches for teaching Religious Education, one traditional (Other RE) and one innovative (Religion Meaning and Life) and their comparative relevance to students in the

W. Sultmann (✉) · J. Lamb
La Salle Academy, Australian Catholic University, Brisbane, Australia
e-mail: William.Sultmann@acu.edu.au

P. Ivers
Faculty of Theology and Philosophy, Australian Catholic University, Brisbane, Australia

M. Craig
University of Divinity, Melbourne, Australia

S. Windsor, O. Franck (eds.), *Intersections of Religion, Education, and a Sustainable World*, https://doi.org/10.1007/978-3-031-81809-7_11

non-compulsory years of schooling in the senior secondary school. The chapter is a response to the challenge for Religious Education in Australia which lies in accommodating the diversity of religious traditions within the school while acknowledging the schools' tradition and purpose and their implications for learning and witness (Queensland Catholic Education Commission [QCEC], 2022). The challenge is made increasingly complex with the declining participation in the Church in Australia (Dixon, 2017) along with the decline but still significant place of Catholic Education within Australia now educating 18.2% of the national population of school students (Independent Schools Australia [ISA], 2021). The challenge is to accommodate RE theory and practice within quality curriculum considering a context characterised by inclusive practices, curriculum preferencing by students, student agency, and diversity in religious affiliations and faith practices. These imperatives increase in significance for students in their final years of schooling where maturation, search for meaning, and competency to operate independently and relationally are paramount. This is no more evidenced than in the field of RE where new pathways are being explored as the reality and significance of RE in the Catholic school remains a priority for the Church in Australia.

11.2 Religious Education

A key determinant of RE curriculum provision is that of context. Education grows out of unique histories, priorities, political circumstances, and legislative frameworks coupled with the changed and changing profiles and expectations of the community. The reminder is provided: "One cannot talk about religious education in the abstract but only about particular forms of religious education, and these forms differ greatly from country to country" (Barnes, 2014, p. 4). Notwithstanding national and regional priorities, RE is fundamentally developed and delivered within the context and culture of Catholic Education:

> The nature of education lies precisely in being able to lay the foundations for peaceful dialogue and allow the encounter between differences with the primary objective of building a better world. It is, first and foremost, an educational process where the search for a peaceful and enriching coexistence is rooted in the broader concept of the human being - in his or her psychological, cultural and spiritual aspects – free from any form of egocentrism and ethnocentrism, but rather in accordance with a notion of integral and transcendent development both of the person and of society (CCE, 2017, para. 15).

The context for Catholic education and more particularly RE holds consistent interest and comment considering global religious, political, social and economic change. For Catholic education, the 'far goal' entails formation of the whole person that is deeply and enduringly humanising (Francis, 2019a). It is a formative experience that goes beyond educating for economic ends to include advancement of human flourishing. "Human flourishing is both the optimal continuing development of human beings' potentials and living well as a human being, which means being engaged in relationships and activities that are meaningful, i.e., aligned with both their own values and humanistic values, in a way that is satisfying to them (ISEEA for UNESCO, 2019, p. 2). Within RE this is realised through experiences that integrate the tradition in support of meaning and purpose for the individual and in contributing to the community. It is to hold in relationship education for meaning and purpose based on the tradition, supportive of evangelisation without ideological colonisation (Francis, 2022b) and being mindful of an age 'awash with information' (Francis 2022c).

> You educate to responsibility, creativity, coexistence, justice, and peace. You educate to the interior life, to be open to the transcendent dimension, to the sense of wonder and contemplation in the face of the mystery of life and creation. You live all this and interpret it in Christ, and translate it into the fullness of humanity (Francis, 2022c, p. 1).

The significance of the Catholic cultural context for Religious Education, and its central place within the overall curriculum of the Catholic school, aligns with the Educational Sustainable Development framework (UNESCO, 2019). The goals projected to 2030 situate learning within the parameters of "advancing policy (integrating ESD in global, regional and national policies of education and sustainable development), learning environments (promoting whole institution approach), teachers and educators (providing capacity development opportunities), youth (providing opportunities for youth engagement) and community (empowering local communities as 'nodal' platforms for all priority action areas)" (UNESCO, 2019, p. 18).

The international experience of RE is identified as being vulnerable in its provision with concerns raised as to its potential collapse, sustainability and marginalisation (Barnes, 2020). Responses are said to rest with two contrasting ways forward. The first is to "attempt to revise the original paradigm... seek its preservation in some form"; and the second; "to admit the problems are insurmountable and be open to the possibility that there may be a new disciplinary model" (Barnes, 2020. p. 3). Barnes goes on to note that "that there is no Archimedean point to view the world of religious education" (p. 6), with either direction needing to be underscored by evidence based and consultative practices. Notwithstanding the selection of directions within diverse contexts, the significance of RE is universal and profound: "to make a positive contribution to the creation of a tolerant, cohesive society where people are respectful of each other and responsible socially" (Barnes, 2014, p. 8).

Conclusions as to what constitutes an integrated, responsive and meaningful RE curriculum reflect the integration of what is educationally sound, socially supported and contextualised by the history and traditions of schools as authentic learning communities. Variation in emphases can be observed across educational spectrums of knowledge to experiential approaches; disciplinary content to student existential concerns; faith-based meaning systems to overall worldviews; and moral foci to ethical considerations. However, what tends to be emphasised with an array of approaches is said to draw from four fundamental principles: student learning; engagement of parents; collaborative partnerships, and responsiveness to diversity, plurality and curriculum development (Braten, 2021).

A challenge for faith-based schools is the provision of RE in once confessional contexts which have now become pluralistic in profile and detraditionalised in nature. The challenge involves balancing long-held religious and educational practices within emerging orientations. The transition is argued to begin with a view of RE as a process as distinct from a product; the teacher as an accompanier who reflects theologically with young people as they dialogue on matters of importance; and, identifying pedagogical practices which support the process of gaining meaning and promote thinking about the transcendent (Roebben, 2021). The shift in direction is not necessarily away from the confessional, the ecclesial or existential, but rather an approach that moves to best practice RE as incorporating these fundamental priorities within frameworks that incorporate inclusion and dialogue. Roebben outlines generative curriculum themes which support this transition: *correlation* – helping students understand their tradition in faith learning; *inclusion* – an appreciation of plurality and the nature of difference among learners; *interpretation* – reflection on experience through applying a common grammar for interpretation; *character education* – incorporating moral education in supporting an ethical perspective on social issues; *narration* – the application of story to inspire reflection and action; *performance* –religious experiences and learning from religious socialisation; and, *spirituality* – being consistent and broad in the application of religious learning.

The opportunity for RE to engage and be informed through encounter and dialogue (Francis, 2013b) provides opportunity and yet demands sensitivity as RE in classrooms are characterised by diversity and carries the implication of RE being in the public sphere (Alexander, 2021). In this regard RE in transition is most ably supported by the integration of key change practices. This is the experience of RE in the European communities which address marginalisation of RE because of increasing pluralisation and secularisation. The elaboration of what becomes critical in transition is elaborated by Skeie (2021) as establishing policy laws (policy); identifying human constants (anthropology); academic rigour; validity and reliability of approaches; contextual stability (consistency in educational practice); and normative choices (values and hopes of the community). In this context the role of RE becomes integral to the mission of the Catholic school.

11.2.1 Religious Education in the Australian Context

International perspectives as to RE being reconceptualised are evident in the Australian context for some time and more recently in curriculum documents issued by Archdioceses such as Melbourne (2018) and Brisbane (2020a). For example, the classroom learning and teaching of religion in the Archdiocese of Brisbane, since 2008, has been characterised by a Reconceptualist Approach; coined by Kieran Scott (1984), to mean changing the way concepts are used for interpreting or explaining something. This educational orientation is one that positions RE to be a robust academic subject in the school curriculum, alongside other academic subjects, with the personal/spiritual dimension accommodated within it. Rossiter (2018), a strong proponent of an educational approach, states:

> Classroom religious education is about educating young people spiritually, ethically and religiously. It is not primarily a religious experience, but it is essentially teaching young people about religion, and about spiritual and moral issues, from a Catholic viewpoint. (p. 1)

Similarly, Scott (1984) has argued in the past that "a reconceptualized religious education takes education as its overarching frame of reference. It self-consciously works out of an educational rather than ministerial framework" (p. 333). He says that "it is the way we go about understanding our own religious tradition, convictions and our God over against the religious identity of 'the other,' the stranger" (p. 333). Building on this further, Gabriel Moran distinguishes between the classroom learning of religion and the practice of religion (Moran, 1981, p. 37). Moran (1981) and Harris & Moran (1992) continue with the perspective that "an education that deserves to be characterised as religious would have to include two quite distinct things: (1) an understanding of religion, starting but not ending with one's own religion; and (2) access to the free and intelligent practice of a particular form of religious life" (Moran, 1981, p. 21).

As part of a larger study (Sultmann et al., 2023), this chapter confines itself to perceptions of senior secondary students who chose to study one of the subjects, Study Religion (SOR) (Queensland Curriculum and Assessment Authority [QCAA], 2019a), *Religion and Ethics* (R&E) (Queensland Curriculum and Assessment Authority [QCAA], 2019b) or Religion, Meaning and Life (RML) (Brisbane Catholic Education [BCE], 2020b) for their RE course. While SOR and R&E are state accredited courses that may contribute toward a tertiary entrance score, RML is a non-accredited RE course that does not contribute to a tertiary entrance score.

11.2.1.1 Religion, Meaning and Life (RML)

In Australia, the history of RE curriculum change across three decades (1990–2020) included the widening of options in RE and the introduction of the Australian Tertiary Admission Rank (ATAR); a system in the state of Queensland for determining entrance to tertiary study. In response to parent and student interests, flexible pathways for studying RE were endorsed in accord with two conditions. First, the affirmation that the Archdiocesan Religious Education Curriculum P-12 (Catholic Education, Archdiocese of Brisbane, 2013) was the authoritative source for curriculum development; and second; under certain circumstances and as required, endorsement was given to the development of flexible delivery options of accredited courses in RE. Against this background of student choice and diversity, approval was given to explore an alternative Archdiocesan non-accredited senior secondary RE course in combination with subjects identified as contributing to tertiary entrance. Nominated as *Religion Meaning and Life* (RML), this course of study would not contribute to a student's tertiary entrance score but would continue to educate and form students "who are challenged to live the gospel of Jesus Christ and who are literate in the Catholic and broader Christian tradition so that they might participate critically and authentically in faith contexts and wider society" (Catholic Education, Archdiocese of Brisbane, 2013, p. 10). In introducing the 2020 RML syllabus, Archbishop Coleridge (2020) noted that the introduction of RML was:

> the beginning of a new era in the way we conceptualise and teach Religious Education. To enter new territory will require imagination, courage and, above all, faith. It will mean letting go of old baggage as we set out, like Abraham, on a new journey led by God. My prayers are with all of you as we set out on the great adventure of opening the infinite treasures of Jesus Christ to the young people whom we are called to serve. (p. 1)

The development of RML curriculum grew from an appreciation that young people are confronted by complexities, dilemmas and conflicting interpretations of life's meaning and purpose. The assumption was that students require, more than ever, critical thinking skills to navigate an uncertain and pluralistic world. The overall goal of RML was for students to experience "an opportunity to access quality theological material in multiple learning modes and engage in a variety of religious experiences and service-learning opportunities" (Brisbane Catholic Education [BCE], 2019, p. 1). Further, the "opportunity for a personal response to key religious ideas through dialogue and an evaluation of a range of secular perspectives" are offered (Brisbane Catholic Education [BCE], 2019, p. 1). In summary, the goals for students entailed the development of: "Twenty-first century skills of critical thinking, creative thinking, communication, collaboration and teamwork, personal and social skills and information and communication technologies skills; and students may develop an ability to engage in an open narrative and dialogue with other religious traditions as well as their own" (Brisbane Catholic Education [BCE], 2019, p. 2). These skills are consistent with the "Aichi-Nagoya Declaration on Education for Sustainable Development states ESD can empower learners to transform themselves and the society they live in by developing knowledge, skills, attitudes, competencies and values required for addressing global citizenship and local contextual challenges of the present and the future, such as critical and systemic thinking, analytical problem-solving, creativity, working collaboratively and making decisions in the face of uncertainty, and understanding the interconnectedness of global challenges and responsibilities emanating from such awareness" (UNESCO, 2019, p. 7).

The implementation of RML commenced in 2019 across two schools. Following the success of the trial course (2019), the course moved into a pilot phase in 2020 where the course was implemented in a further 11 schools. During 2021 the number of pilot schools grew to 17.

11.2.1.2 Traditional Approaches: Religion and Ethics and Study of Religion

Religion and Ethics is an applied senior syllabus. In Queensland, Australia, an Applied subject or course of study is suited to students who are primarily interested in pathways beyond senior secondary schooling that lead to a student's ATAR. As an Applied subject, *Religion and Ethics* provides opportunities for students to explore their personal values and life choices and the ways in which these are related to their beliefs. Religion is viewed as a faith tradition based on a common understanding of beliefs and practices. The term ethics is argued to be a system of moral principles and approaches to making decisions for the good of the individual and society. The course, *Religion and Ethics* focuses on the personal, relational and spiritual perspectives of the individual. It supports students to investigate and critically reflect on the role and function of religion and ethics in the society and world in which they live (QCAA, 2019b).

Study of Religion, as a general course of study is one that is suited to students who are interested in pathways beyond senior secondary schooling that lead to tertiary studies and to pathways for vocational education and training and work. As a general subject, *Study of Religion* provides opportunities for students to investigate religious traditions and how they have influenced, and continues to influence, people's lives (QCAA, 2019a). Furthermore, students become aware of their own religious beliefs, the religious beliefs of others, and how people holding such beliefs are able to co-exist in a pluralist society (QCAA, 2019a). In this subject, students engage with an inquiry approach to learning (QCAA, 2019a).

11.2.1.3 Research Question and Sub-Questions

Do RML students differ in their overall level of satisfaction with their study compared to students who study other RE curriculum options?

1. What topics hold most significance in RE?
2. What pedagogical approaches are significant?
3. What outcomes are of most significance?

11.3 Method

11.3.1 Participants

11.3.1.1 School Communities

Data were collected from Year 11 and 12 RE students (*Religion Meaning and Life* [RML] *Study of Religion* [SoR] and *Religion and Ethics* [R&E]), from 17 pilot RML schools. The school communities that participated in the study included nine Brisbane Catholic Education (BCE) and eight Religious Institute or Ministerial PJP Authorities.

11.3.1.2 Students

Three hundred and fifty-three students participated in the study with more female than male students. Forty-two percent of participants were studying RML; 81% of all participants elected an ATAR pathway used to determine tertiary entrance.

Table 11.1 Student characteristics (N = 353)

Characteristics	Categories and distribution		
Gender	Male: 134 (38%)	Female:211 (59.8%)	Other:8 (2.3%)
Pathway	ATAR: 286 (81%)	Non-ATAR: 40 (11.3%)	Mixed: 27 (7.6%)
RE course	Other RE: 203 (57.5%)	RML: 150 (42.5%)	
School type	Secondary: 328 (92.9%)	P-12: 25 (7.1%)	
School size	<500: 33 (9.3%)	>500: 320 (90.7%)	
Location	Metropolitan: 199 (56.4%)	Regional/rural: 154 (43.6%)	

Students were from metropolitan and rural/remote locations and more than 90% were in schools larger than 500 students (see Table 11.1).

11.3.2 Data Collection

11.3.2.1 Process

Participant information letters outlining the project and its expectations were sent to the principals of the 17 pilot schools. Communications outlined survey goals, general content, timing and completion procedures, confirmed anonymity and provided digital links to the survey. These letters were distributed to the students and their parents.

11.3.2.2 Survey Instruments

The online survey was developed and made available through the Qualtrics platform. Survey questions focussed on three themes: RE topics; pedagogical approaches; and outcomes. In addition, a measure of the overall level of satisfaction was sought from all students undertaking RE in Years 11 and 12.

Each survey commenced with a series of demographic questions relevant to student enrolment. Survey questions were framed from four sources: *Framing Paper: Religious Education in Australian Catholic schools* (NCEC, 2018); the RML Pilot 2020 Course Handbook Version 2 (BCE, 2020b); case study examples from recent research (Sultmann et al., 2020, 2022) and findings of a national colloquium on ways forward in RE (Hall & Sultmann, 2020). Participants were asked to rate survey statements through application of a Likert scale using a range from 1 to 5: where 1 was the lowest score; and 5 was the highest score.

11.3.3 Data Analysis

11.3.3.1 Survey

Data analysis employed descriptive statistics and are presented in tabular form with percentage responses, frequencies and means. In addition, a series of *t*-tests were conducted to assess difference in the levels of satisfaction between students studying RML and students studying other RE courses.

11.4 Results

11.4.1 Research Question

The overall research question examined the difference between RML students and Other RE students as to their overall level of satisfaction with their elected RE course.

Results indicate a consistency in the level of satisfaction of RE across the curriculum options within the pilot schools. This was evident from overall means and in the distribution of scores (see Table 11.2).

11.4.1.1 Sub-Research Question 1

The first sub-question examined the topics that held most significance in RE. Ethics and Other World Religions featured most prominently in content priorities across all RE options. Notably traditional content areas of Christian religious beliefs, Scripture, Jesus, Church and Prayer all were rated as low (see Table 11.3). Comparison of mean scores across the content themes of Jesus and Other World Religions were statistically significant and higher for those studying RML (see Table 11.4). Notwithstanding, overall mean scores were modest and were at or slightly above the mid-range.

Table 11.2 Evaluation of RE satisfaction according to RE course

	% Response Proportions					
RE Options	1	2	3	4	5	Mean
Other RE	18.2	12.8	25.6	29.1	14.3	3.1
RML	13.3	14.7	27.3	26.0	18.7	3.2

Table 11.3 Importance of RE content by topic according to RE course

Topics	Course	% Response Proportions					Mean
		1	2	3	4	5	
Christian religious beliefs	Other RE	26.1	20.7	26.1	18.7	8.4	2.6
	RML	21.3	20.0	28.0	19.3	11.3	2.8
Scripture	Other RE	35.0	19.2	30.5	10.3	4.9	2.3
	RML	28.0	24.0	29.3	11.3	7.3	2.5
Jesus	Other RE	32.5	19.2	29.1	13.3	5.9	2.4
	RML	27.3	15.3	22.7	19.3	15.3	2.8
Catholic social teaching	Other RE	25.1	13.3	27.6	23.2	10.8	2.8
	RML	21.3	14.0	28.0	19.3	17.3	3.0
Christian traditions	Other RE	29.6	17.2	29.1	16.7	7.4	2.6
	RML	23.3	20.0	35.3	14.0	7.3	2.6
Other world religions	Other RE	25.1	13.8	22.7	19.2	19.2	2.9
	RML	15.3	9.3	22.0	24.0	29.3	3.4
Church	Other RE	34.5	25.1	26.1	10.8	3.4	2.2
	RML	27.3	25.3	32.0	9.3	6.0	2.4
Prayer	Other RE	34.0	21.2	28.6	11.3	4.9	2.3
	RML	28.7	18.0	30.0	15.3	8.0	2.6
Sacraments	Other RE	36.5	21.7	26.1	11.8	3.9	2.3
	RML	30.0	24.7	28.0	10.0	7.3	2.4
Doctrines (Teachings of the Church)	Other RE	35.5	19.7	24.6	11.3	8.9	2.4
	RML	30.7	18.7	24.7	19.3	6.7	2.5
Christian life	Other RE	34.0	23.2	25.1	13.8	3.9	2.3
	RML	24.7	28.0	26.0	12.0	9.3	2.5
Ethics	Other RE	15.3	8.9	20.7	18.2	36.9	3.5
	RML	12.7	8.0	19.3	23.3	36.7	3.6

Table 11.4 Independent samples t-test for importance of RE content by topic

Topics	Mean	Standard Deviation	df	t	Sig (2-tailed)	Cohen's d
Jesus						
Other RE	2.4	1.23	293	−2.8	0.007	0.33
RML	2.8	1.42				
Other world religions						
Other RE	2.9	1.45	351	−3.2	0.002	0.35

11.4.1.2 Sub-Research Question 2

The second sub-question examined the significance of pedagogical approaches. Pedagogical approaches which included dialogue, use of media and technology, and solving real life problems featured positively and consistently across all RE course options. Notably, the least favoured pedagogical approaches reported by students were praying and the use of a textbook (see Table 11.5). Dialogue was appreciated

Table 11.5 Student perceptions of pedagogical approaches to learning RE

		% Response Proportions					
Criteria	Course	1	2	3	4	5	Mean
Dialogue	Other RE	17.7	11.3	24.6	24.6	21.7	3.2
	RML	8.7	9.3	14.0	35.3	32.7	3.7
Textbook	Other RE	39.4	18.2	26.1	12.8	3.4	2.2
	RML	32.7	30.7	24.0	10.7	2.0	2.2
Applying scripture	Other RE	34.5	19.7	26.1	12.3	7.4	2.4
	RML	28.0	27.3	26.0	14.7	4.0	2.4
Media and technology	Other RE	14.3	8.9	30.0	26.6	20.2	3.3
	RML	8.0	9.3	26.7	35.3	20.7	3.5
Teaching team (More than one teacher)	Other RE	35.0	22.2	28.6	8.9	5.4	2.3
	RML	24.7	26.7	27.3	16.0	5.3	2.5
Team teaching (Peer to peer learning)	Other RE	21.2	12.8	25.1	26.1	14.8	3.0
	RML	16.7	8.7	28.7	24.0	22.0	3.3
Projects	Other RE	20.2	17.2	27.1	24.1	11.3	2.9
	RML	20.7	22.0	31.3	16.0	10.0	2.7
Applying RE	Other RE	21.2	10.8	24.1	21.7	22.2	3.1
	RML	18.0	18.0	27.3	17.3	19.3	3.0
Praying	Other RE	53.7	17.2	16.7	8.9	3.4	1.9
	RML	42.7	26.0	17.3	8.0	6.0	2.1
Solving real life problems	Other RE	18.2	6.9	26.1	18.2	30.5	3.4
	RML	8.7	12.7	24.0	26.7	28.0	3.5

Table 11.6 Independent samples t-test importance of pedagogical approaches to learn RE

Topics	Mean	Standard Deviation	df	t	Sig (2-tailed)	Cohen's d
Dialogue						
Other RE	3.2	1.38	336	−3.8	0.000	0.37
RML	3.7	1.25				

by all students however, a significant difference in perception with respect to the value of dialogue was reported by RML students (see Table 11.6).

11.4.1.3 Sub-Research Question 3

The third sub-question examined which outcomes were of most significance?

Nominated curriculum outcomes for all RE students were generally rated as being achieved except for Praying (see Table 11.7). However, analyses of outcome differences for students across RML and those pursuing Other RE options identified some outcomes as being significantly higher for RML students. These included: Listening to others, Dialoguing, Making connections, Understanding myself, and Contributing to community (see Table 11.8).

Table 11.7 Student perception of outcomes

Outcomes	Course	% Response Proportions 1	2	3	4	5	Mean
Listening to others	Other RE	22.2	10.8	31.0	26.1	9.9	2.9
	RML	11.3	10.0	29.3	30.7	18.7	3.4
Dialoguing with others	Other RE	21.7	12.3	26.6	28.1	11.3	3.0
	RML	10.7	12.0	26.0	34.7	16.7	3.3
Critiquing information	Other RE	21.7	13.8	30.5	19.2	14.8	2.9
	RML	13.3	13.3	36.0	21.3	16.0	3.1
Making connections	Other RE	22.7	12.3	27.1	27.6	10.3	2.9
	RML	10.7	12.0	33.3	27.3	16.7	3.3
Understanding myself	Other RE	29.1	14.3	28.1	21.7	6.9	2.6
	RML	19.3	12.0	30.7	22.7	15.3	3.0
Contributing/community	Other RE	26.6	17.7	26.6	21.2	7.9	2.7
	RML	13.3	13.3	26.0	26.7	20.7	3.3
Applying teachings	Other RE	30.0	15.3	26.1	21.2	7.4	2.6
	RML	24.7	20.7	26.0	16.7	12.0	2.7
Praying	Other RE	54.7	18.2	13.8	10.3	3.0	1.9
	RML	44.0	23.3	19.3	8.0	5.3	2.1

Table 11.8 Independent samples t-test for outcomes of RE study

Topics	Mean	Standard Deviation	df	t	Sig (2-tailed)	Cohen's d
Listening to others						
Other RE	2.4	2.9	351	−3.3	0.001	0.12
RML	2.8	3.4				
Dialoguing with others						
Other RE	2.9	3.0	351	−2.9	0.004	0.15
RML	3.4	3.3				
Making connections						
Other RE	2.9	1.3	351	2.7	0.007	0.33
RML	3.3	1.1				
Understanding myself						
Other RE	2.6	1.2	351	2.8	0.005	0.32
RML	3.0	1.3				
Contributing to community						
Other RE	2.7	1.3	351	4.4	0.000	0.46
RML	3.3	1.3				

11.5 Discussion

The discussion offers an initial summary of the research findings. Second, an interpretation of these research findings for students studying traditional academic approaches, in comparison to a more applied experiential approach such as in RML, is then developed. Third, the discussion identifies the implications of the research for RE at the senior secondary level.

Satisfaction with RE. Overall, the comparison of perceptions by students studying academic approaches with an innovative reconceptualist applied approach indicated that there was no difference in the level of student satisfaction.

Topics of Significance. For topics studied, the general level of appreciation of Ethics and Other World Religions were prioritised across all RE courses. However, for RML students there was a statistically significant and positive difference for the topics of Jesus and Other World Religions.

Pedagogy of significance. Pedagogy based on dialogue, media and technology together with solving real world life problems were consistently valued. The importance of Dialogue in RML pedagogy, in comparison to other courses of study, was significantly different and positive. Pedagogical approaches which include and advance interpersonal behaviours and engage community were established as significantly different and positive for RML students.

Outcomes of RE. Outcomes nominated for RML and Other RE courses were achieved. However, five of the eight were statistically significant and positively different for RML students. These findings warrant further discussion given that the overall satisfaction was similar for both groups.

The topics of Jesus and Other World Religions were statistically significant and higher for those in the RML group. The RML students had an intentional focus on the application of the Gospel both theoretically and practically from their field-based learning experiences. In contrast, the academic focus in the Other RE courses covered a diversity of religious traditions and were formally taught within a classroom experience. Notably, the finding as to the significance of dialogue was valued by both groups. Nonetheless, the importance of dialogue for those in the RML group was significantly different and positive than for those in the Other RE courses. Clearly, dialogue was important as a pedagogical approach which sought to build understanding of concepts within the designated unit of study. However, the concept of dialogue for the RML group not only facilitated understanding but also contributed to understanding within an applied context with a focus on personal development and integration. Moreover, the RML experience underscored the agency of the student in applying and integrating the learning experience. In such instances, evangelisation takes on immediacy, relevance, encounter and potential engagement.

> Catholic schools remain essential places for the evangelization … These include a fresh experience of the kerygma, wide-ranging dialogue, interdisciplinary and cross-disciplinary approaches, the promotion of a culture of encounter, the urgency of creating networks and an option in favour of those who are least, those whom society discards. Similarly important is the ability to integrate the knowledge of head, heart and hands. (Francis [CV], 2019a, para. 222)

The significant Outcomes for the RML group included: listening to others; dialogue; making connections; understanding self; and contributing to community. These presented as flowing naturally from the combination of theory and practice within the RML course and are consistent with the ESD goals (UNESCO, 2019) and demonstrate a way to promote these goals (O'Flaherty & Liddy, 2018). These outcomes aligned with the essential character of the RML course which included: a learning environment which encouraged curiosity and reflection through the use of non-judgemental language; freedom to critique the relevance of religion; capacity to translate theory into practice from experiential service based learning; the value of learning in small groups which maximised peer to peer engagement and student to teacher relationship; and the flexibility of the teacher to act as "Specialist, Moderator and Witness" as described by Pollefeyt (2008, p. 15). A combination of these factors provided a dynamism in the course that reflected its aspirations and were witnessed in its delivery.

Third, the findings of the study provide implications for the delivery of RE more broadly in the senior secondary curriculum. The Catholic school in Australia is 'two hundred years young' (Catholic Bishops of Australian, 2021); demonstrating across two millennia change and growth in relationship to its religious, social and cultural heritage within the community it serves. Notwithstanding the revelance of the study within the Australian context, the Catholic school on the threshold of the new millennium articulates this contemporary relationship for Catholic schools generally as the home and school of communion (CCE, 2014); an image that speaks to the diversity and religious plurality of its community, and the challenge of its formative educational project as providing a basis for fraternal humanism (CCE, 2017). The focus is one of expressing and witnessing to the Spirit of Christ in curricula which is "rich in humanity and capable of being with young people in a style of pedagogy that helps promote their human and spiritual growth" (Francis, 2014, para. 6). This breadth of application draws upon generic principles for both traditional and innovative curriculum which give emphasis to faith traditions, dialogical processes, experiential learning, the integral and shared agency of teacher and student and applications within ethical frameworks supportive of the development of personal meaning systems.

In summary, student ratings of RE identified satisfaction with all RE courses. The suite of RE options, Other RE and RML, presented as respectful of learner priorities, differences and needs. Questions of inclusion, diversity, dialogical processes were able to be addressed in providing relevant and flexible curriculum: "helping to develop mature people who are simple, competent and honest, who know how to love with fidelity, who can live life as a response to God's call and their future profession as a service to society" (CCE, 2014, para. 8). In this context, Religious Education "expands students' spiritual awareness and religious identity, fostering the capacities and skills of discerning, interpreting, thinking critically, seeking truth and making meaning. It challenges and inspires their service to others and engagement in the Church and the world" (NCEC, 2018, p. 7). Ethics within RE was found to be significant in this regard as was the relevance of Other World Religions. This trend reflects directions internationally in RE provision being more responsive to pluralism, change and the importance of ethical and experiential learning. Notwithstanding, this research identified the significance of Jesus and Other World Religions as a basis for ethical applications.

The following common elements present as significant to both Other RE and reconceptualist courses such as RML for RE in the senior secondary school.

- Appreciation of the Catholic Christian meaning system and Other World Religions as a basis for ethical applications.
- Advancing dialogical skills in listening to others, making connections, understanding self, and contributing to community in the teaching and learning process.
- Promoting the role of teacher as specialist, moderator and witness in establishing the space for inclusion, respectful dialogue and nurturing of relationships.
- Aligning experiential learning approaches with unit content and pedagogy.
- Pursuing continuous research as existing and innovative RE courses are subject to development in response to changed and changing Catholic school communities.

This study evaluated an innovative RE course and compared it with existing RE options for students in the senior secondary school. More particularly, the development and review of RML was responsive to a changed and changing Catholic school context while being faithful to its tradition and responsive to parent choice in RE. The research entailed an opportunity to identify matters of content, pedagogy and outcomes of significance and contrast these with similar domains in existing academic RE curricula. The findings confirm factors of engagement, curriculum quality and provision, staff engagement, performance monitoring and interpretation and a commitment to policy imperatives transcendent.

The research has relevance within and beyond Australia when attempting to position Catholic schools and RE within plural and multicultural contexts. It draws from empirical research as a basis for 'what works' while offering a unique approach to understanding and educating within a distinctive faith based educational philosophy. At the same time, it reinforces the importance of educational sustainable development goals applicable across the whole curriculum.

References

Alexander, H. (2021). Presentation abstract: Religious Education in the public sphere. John Hull memorial lecture. In *International seminar on Religious Education and values. Digital conference*, 26–30 July, Gothenburg, Sweden.

Barnes, L. P. (2014). *Education, religion and diversity*. Routledge.

Barnes, L. P. (2020). *Crisis, controversy and the future of Religious Education*. Routledge.

Braten, O. (2021). MOVe with RE-searchers across the curriculum and national boarders to achieve deep learning: Experimenting with teaching methodology developed in Exeter, UK, in A local "university school" in Trondheim, Norway. Presentation abstract: Facing the unknown future: Religion and education on the move. In *International seminar on Religious Education and values. Digital conference*, 26–30 July, Gothenburg, Sweden.

Brisbane Catholic Education. (2019). *Religion meaning and life course handbook*. Brisbane.

Brisbane Catholic Education. (2020a). *Religious Education curriculum P-12*. https://curriculum.bne.catholic.edu.au/Curriculum/LearningArea?learningAreaName=Religious%20Education

Brisbane Catholic Education. (2020b). *Religion meaning and life pilot 2020 course handbook, Version 2*. Brisbane.

Catholic Bishops of Australian. (2021). *200 years young*. https://www.ncec.catholic.edu.au/images/Pastoral_Letter_200_Years_Young_18022021.pdf

Catholic Education Melbourne. (2018). *Religious Education curriculum framework documentation (Draft)*. Archdiocese of Melbourne.

Coleridge, M. (2020). *The Shape of Religious Education: Letter from the Archbishop of Brisbane*. http://www.rec.bne.catholic.edu.au/The%20Shape%20of%20Religious%20Education/Pages/Letter-from-the-Archbishop-of-Brisbane.aspx

Congregation for Catholic Education [CCE]. (2009). *Circular letter to the presidents of Bishops' conferences on Religious Education in schools*. Retrieved from https://www.vatican.va/roman_curia/congregations/ccatheduc/documents/rc_con_ccatheduc_doc_20090505_circ-insegn-relig_en.html

Congregation for Catholic Education [CCE]. (2013). *Educating to intercultural dialogue in Catholic schools: Living in harmony for a civilization of love*.

Congregation for Catholic Education [CCE]. (2014). *Educating today and tomorrow: A renewing passion*. http://www.vatican.va/roman_curia/congregations/ccatheduc/documents/rc_con_ccatheduc_doc_20140407_educare-oggi-e-domani_en.html

Congregation for Catholic Education [CCE]. (2017). *Educating to fraternal humanism: Building a civilization of love 50 years after Populorum Progressio*.

Dixon, R. (2017). The Australian Church in 2030: What the research predicts. *St Thomas More Forum. Catholica*. Retrieved from https://www.catholica.com.au/gc2/occ4/152_occ4_030517.php

Francis. (2013b). *The joy of the Gospel: Evangelii Gaudium [EG]*. https://www.vatican.va/content/francesco/en/apost_exhortations/documents/papa-francesco_esortazione-ap_20131124_evangelii-gaudium.html

Francis. (2014). *Address of Pope Francis to participants in the plenary session of the Congregation for Catholic Education*. Retrieved from: https://w2.vatican.va/content/francesco/en/speeches/2014/february

Francis. (2019a). *Christus vivit*. www.vatican.va/content/francesco/en/apost_exhortations/documents/papa-francesco_esortazione-ap_20190325_christus-vivit.html

Francis. (2022b). *Vatican News: Pope wars Catholic educators against ideological colonization*. http://vaticannews.va/en/pope/news/2022-11/pope-warns-catholic-educators-against-ideological-colonization.html

Francis. (2022c). *Vatican News: Hope to De La Salle Brothers: World needs a new education pact*. www.vaticannews.va/en/pope/news/2022-05/pope-de-la-salle-brothers-world-needs-new-pact-on-education.html

Hall, D., & Sultmann, W. F. (2020). Ways forward in Religious Education: Reflections of an Australian colloquium. *eJournal of Catholic Education in Australasia, 4*(1), 1–23.

Harris, M., & Moran, G. (1992). Catechetical language and Religious Education. [In English]. *Theology Today, 49*(1), 21.

Independent Schools Australia. (2021). School enrolment trends and projections. *Independent Schools Association (ISA)*.

ISEEA for UNESCO. (2019). *Meaning(s) of human flourishing and education*. http://en.unesco.org/futuresofeducation/sites/default/files/2021-03/Flourishing%20and%20Education_ISEEA%20Research%20Brief.pdf

Moran, G. (1981). *Interplay: A theory of religion and education*. St Mary's PR.

National Catholic Education Commission. (2018). *Framing paper: Religious Education in Australian Catholic Schools*. https://www.ncec.catholic.edu.au/images/NCEC_Framing_Paper_Religious_Education.pdf.com_content&view=article&id=16

O'Flaherty, J., & Liddy, M. (2018). The impact of development education and education for sustainable development interventions: A synthesis of the research. *Environmental Education Research, 24*(7), 1031–1049. https://doi.org/10.1080/13504622.2017.1392484

Pollefeyt, D. (2008). *Difference matters. A hermeneutic-communicative concept of didactics of religion in a European multi-religious context*. Retrieved from Difference Matters (kuleuven.be).
Queensland Catholic Education Commission (QCEC). (2022). *Mission in Practice: A lens for examining Catholic school identity*. ACU_Mission-Practice-Framework_SinglePg.pdf (qcec.catholic.edu.au)
Queensland Curriculum and Assessment Authority (QCAA). (2019a). *Study of religion v1.2*. https://www.qcaa.qld.edu.au/senior/senior-subjects/humanities-social-sciences/study-of-religion/syllabus
Queensland Curriculum and Assessment Authority (QCAA). (2019b). *Religion and ethics*. https://www.qcaa.qld.edu.au/senior/senior-subjects/humanities-social-sciences/religion-ethics
Roebben, B. (2021). Religious educational leadership in times of upheaval. How to build sustainably on insights from the past? Abstract presentation. In *International seminar on Religious Education and values. Digital conference*, 26–30 July, Gothenburg, Sweden.
Rossiter, G. (2018). *Life to the full – The changing landscape of contemporary spirituality: Implications for Catholic School Religious Education*. Agora.
Scott, K. (1984). Three traditions of Religious Education. *Religious Education, 79*(3), 323–339. https://doi.org/10.1080/0034408400790302
Skeie, G. (2021). Presentation abstract: Who cares about RE: Recent curriculum development in Norwegian RE in the context of change in educational policy and the role of religion in society. In *International seminar on Religious Education and values. Digital conference*, 26–30 July, Gothenburg, Sweden
Sultmann, W. F., Lamb, J., Hall, D., & Borg, G. (2020). Student perceptions of a trial Religious Education curriculum: Establishing baseline data. *Journal of Beliefs and Values*. https://doi.org/10.1080/13617672.2020.1818923
Sultmann, W. F., Lamb, J., Hall, D., & Borg, G. (2022). Teacher and student evaluations of a trial RE curriculum: Implications for scale. *British Journal of Religious Education*. https://doi.org/10.1080/01416200.2022.2042191
Sultmann, W., Lamb, J., Ivers, P., & Craig, M. (2023). *Re-imagining senior secondary Religious Education: Evaluating the religion meaning and life curriculum* (SpringerBriefs in education). Springer.
UNESCO. (2019). *Framework for the implementation of Education for Sustainable Development (ESD) beyond 2019*. http://unesdoc.unesco.org/ark:/48223/pf0000370215

William Sultmann AM, PhD is an Associate Professor at Australian Catholic University. He is a teacher, psychologist, practical theologian and leader with experience in education, health and welfare across Government and Catholic sectors. Current responsibilities include university teaching, higher degree teaching and research, and community engagement in governance boards.

Janeen Lamb PhD is an Associate Professor at Australian Catholic University with expertise in Educational Leadership, Curriculum and Pedagogy, Statistical applications and Research Methods. Professional experience and expertise have been evidenced in positions as Director, Higher Degree Research in Education, and in instructional leadership in teaching doctoral courses.

Peter Ivers EdD is a Senior Lecturer and National Coordinator for Postgraduate Studies in the Faculty of Theology and Philosophy at Australian Catholic University. He has extensive experience in a variety of leadership roles in different educational settings. His doctoral work investigated the emergence of educational discourses on Religious Education.

Mark Craig PhD candidate and a secondary school teacher of religion, has worked in full time ministry with the Archdiocese of Brisbane, and is currently employed as an Education Officer with Brisbane Catholic Education. He is completing a PhD in Philosophy at the University of Divinity, Melbourne.

Chapter 12
Rethinking Sustainability: Contributions of Religious Literacy Education

Bruce Grelle

12.1 Introduction

It has become common to think about "sustainability" in terms of three overlapping components known as the 3 Ps of *planet*, *profit*, and *people* or as the 3 Es of *environment*, *economy*, and *equity*. The idea here is that the long-term well-being of our natural environment, our economies, and our societies are interconnected. For life of any sort to thrive we need clean air and water and a liveable climate. For human life to thrive we need economies capable of providing for the material necessities of life, and we need societies that exhibit some degree of safety, stability, and fairness.

Few people deny that education in the natural and social sciences has a bearing on questions of sustainability. We cannot understand the greenhouse effect and the impacts of climate change without some basic education in the physical and biological sciences. Likewise, we cannot understand problems of power and inequality or struggles over public policy without some familiarity with concepts drawn from economics, sociology, and political science.

This chapter argues that religious literacy education is just as basic to understanding the crisis of sustainability in which we find ourselves. Through study of diverse religious and secular worldviews we learn to recognize the ambiguous moral and environmental implications of the deep-seated ideas and taken-for-granted assumptions that shape our individual and societal lifestyles. This chapter explores ideas and assumptions that underlie the economic and technocratic perspectives that dominate current discussions of sustainability, and it identifies components of some more ecologically minded alternatives.

B. Grelle (✉)
Professor Emeritus in the Department of Comparative Religion and Humanities,
California State University, Chico, CA, USA
e-mail: BGrelle@csuchico.edu

S. Windsor, O. Franck (eds.), *Intersections of Religion, Education, and a Sustainable World*, https://doi.org/10.1007/978-3-031-81809-7_12

12.2 Sustainability

We say that something is sustainable if it can be maintained over a period of time. Contemporary discussions typically assume that what we are trying to maintain is the way of life that most of us in the Global North have come to take for granted as normal, a way of life that millions of people throughout the world are also aspiring to emulate.

Yet as Canadian ecologist William Rees (2014) reminds us, the industrial and technological developments that have made our consumer lifestyle possible have taken place in just the last few hundred years. That's a mere blip of time in the vast span of millennia that human beings have lived on this planet from prehistory to the present. What we take for granted as normal is dramatically unlike the way that humans have lived at any other point in history.

For those of us fortunate enough to participate, it is a lifestyle of material abundance, comfort, and convenience. But it is also a way of life that is depleting nature's resources and producing pollution at a rate and in an amount that far outstrips the ability of natural systems to replenish themselves.

Rees and colleagues developed the concept of the "ecological footprint" in the 1990s (Wackernagel & Rees, 1996). It has become a helpful and widely-used tool for educating about the *unsustainability* of the fossil fuel-based consumer economy that continues to spread around the world and about the inequalities in resource consumption between and within countries. The ecological footprint measures human impacts on the planet by calculating the amount of land and water area that is required to provide the resources consumed, and to absorb the waste produced, by an individual's or group's way of life given the prevailing technology. In order to be sustainable, our ecological footprint must be less than what can be naturally replenished and absorbed by the Earth's natural systems (Heinberg, 2012, p. 2).

'Earth Overshoot Day' marks the date on which humanity's demand for nature's resources and services in a given year exceeds what can be regenerated in that year. In 2022 Earth Overshoot Day occurred on July 28. In just a little under 7 months' time, human use of natural resources and services had exceeded what can be regenerated by the Earth in 12 months (Global Footprint Network, 2023). Moreover, this state of overshoot results mainly from meeting the material demands of just the wealthiest fifth of the human population who consume roughly 70% of the world's economic output, with the poorest fifth surviving on only 2% (Rees, 2015, p. 6).

To be sure, previous civilizations – Sumerians, Mayans, and others – have overshot the ability of their local ecosystems to sustain themselves and have collapsed as a result (Brown, 2009, p. 3). What makes the current situation unique is that this time the scope of the collapse threatens to be global rather than local. Ecological and social crises facing the present world order are interconnected and far-reaching – as extreme weather events, deforestation, ocean acidification, mass extinctions, melting glaciers, and rising sea levels are projected to bring about widespread famine, proliferating disease, millions of climate refugees, and increased political instability over the next few decades (Robinson, 2014, pp. 231–232).

What does all this have to do with teaching and learning about religions and beliefs?

William Rees thinks that one of the main barriers to recognizing the *un*sustainability of our present global political economy is the widespread influence of mainstream economic theory (2015). With its emphasis on perpetual economic growth, mainstream economics promotes an outlook that is blind to the limits of the biophysical world. It imagines a world of endless expansion – more people, more energy, more technology, more abundance (Jackson & Jensen, 2022, p. 91). This way of looking at the world is deeply entrenched among economists, politicians, business leaders, journalists, and everyday citizens. But it is an outlook that is incapable of fully recognizing, let alone resolving, the crisis of sustainability in which we find ourselves.

A growing number of scholars have argued that this economic way of looking at the world, along with the practices and institutions that it authorizes, amounts to a new secular religion of sorts, a system of faith that has become dominant in today's world (Foltz, 2007). Helping students and the general public to learn more about this view of life and its far-reaching influence, and helping them to recognize and envision possible alternatives, is a main contribution that education about religions and beliefs can make to contemporary discussions of environmental, economic, and social sustainability.

12.3 Economic Religion

According to the neoclassical model of economic behavior favored by the majority of professional economists and promoted in schools, individuals are primarily motivated by the pursuit of self-interest (Council for Economic Education, 2010; Nord, 2010, pp. 47–52, 228–231; Nord & Haynes, 1998, pp. 105–114). The chief purpose of the economy is to maximize utility for individuals and profits for businesses. This is best achieved when markets are allowed to operate with as little interference as possible. Free markets promote competition, productivity, and growth, which benefit society as a whole.

No doubt, the neoclassical model – the majority paradigm – presents many shrewd insights into the wellsprings of human behavior. It provides a compelling framework for analyzing and organizing human societies. But the historical contingency of this framework – its origins in the eighteenth century European Enlightenment – and the fact that it is one particular picture of human life among other possible interpretations, frequently goes unacknowledged.

When economics is taught in schools and discussed in mainstream media it is typically presented as a neutral scientific description of the way the world works. (Rees, 2015; Nord, 2010, pp. 49–52; Moore, 2007, pp. 171–173). Yet as economist Robert H. Nelson has argued, mainstream economic theory is itself a system of belief with powerful moral and ecological implications. He describes it as a "religion" in a broad sense insofar as it provides a framework of values and a sense of human purpose that is rooted in a particular view of self and society. It values money, competition, and individualism. It promises salvation through unlimited economic growth and a rising GDP (Nelson, 2010b; see also 2001, 2010a). Its basic premises find popular expression in the notions that the more you have, the happier you will be; whoever dies with the most toys wins!

Nelson is convinced that people must have some form of religion; no one can live entirely free from a framework of meaning, though of course not all religions require a God, as Judaism and Christianity do (Nelson, 2010b). Others have argued that economic religion does indeed have gods that it worships and in which it places its faith. Protestant theologian Harvey Cox (1999, 2016) has described the rise of an increasingly influential "business theology" that revolves around the idea of "The Market as God," and he argues that faith in The Market has become one of the main competitors with the world's traditional religions.

Likewise, Buddhist philosopher David Loy (1997) has claimed that "market capitalism … has already become the most successful religion of all time, winning more converts more quickly than any previous belief system or value-system in human history" (p. 276). He bases this claim on a functionalist definition of religion as that which "grounds us by teaching what the world is, and what our role in the world is" (p. 275). Loy contends that science and consumerism have increasingly supplanted the role of traditional religions in peoples' lives. With the market as its god, economists as its theologians, and the advertising industry as its missionary vanguard, the "religion" of market capitalism promises salvation through technological progress and ever-increasing production and consumption of material goods and services.

Like a god, the "invisible hand" of the market metes out rewards or punishments according to our obedience or disobedience to its laws (of competition, supply and demand, etc.). The market must be appeased or else its wrath will descend upon entire nations. This god-like status is revealed in such oracular news headlines as "The Market was Happy with News of the Latest Jobs Report," or "The Market Responded with Displeasure to News of the Latest Financial Bailout Request" (Mohr, 2017).

Others have identified Growth as the god of economic religion. Process theologian John Cobb has argued that in the twentieth century "neoclassical economics became the theology of those who saw economic growth as the savior of humankind." While allowing that it "does not dominate the spirituality of all peoples, it is [nonetheless] the 'religion' that governs planetary affairs" and is rooted in a righteous "devotion" to the "increase of economic production" (cited in Nelson, 2010a, p. 1). Likewise, environmental historian J.R. McNeill writes that during the twentieth century "a more flexible and seductive religion succeeded where communism failed: the quest for economic growth. Capitalists, nationalists – indeed almost everyone … – worshipped at this same altar" (cited in Nelson, 2010a, p. 1).

Whether it is the God of Growth or the Market that is most central to the theology of economic religion, it is clear that we are dealing here with a widespread and influential way of construing life. Yet we should pause to ask whether "economic religion" is really a religion in anything more than a metaphorical sense? Do the beliefs and values associated with global capitalism really amount to a new world faith system?

If we accept the functional definition of religion proposed by Nelson, Loy, and others; if we agree that "faith" is about being loyal to and trusting in a set of beliefs and values that orient the entirety of one's life, then we might well agree that aspects of global economic life deserve to be described as religious. However, even if we do not accept these definitions and want to maintain a sharper distinction between "religious" and "secular" perspectives and traditions, I think we can still agree that we are dealing here with a distinctive *worldview* and *ethos* and that approaching these from a religious studies perspective can be instructive.

Such a claim is nothing new. Ninian Smart's pioneering work famously argued that religious studies should include attention to secular ideologies such as nationalism, Marxism, scientism, and humanism, and that these worldviews exhibit doctrinal, mythical, ritual, and ethical dimensions that parallel those of traditional religions (1995). Likewise, documents such as the *Toledo Guiding Principles on Teaching about Religions and Beliefs in Public Schools* (OSCE, 2007) and *Signposts – Policy and Practice for Teaching about Religions and Non-Religious World Views in Intercultural Education* stress that education in state schools should not only include attention to traditional religions but also to "non-religious convictions and world views" (Jackson, 2014, pp. 67–76). John Valk (2021) has explored parallels between "religious," "spiritual," and "secular" worldviews, including consumerism, and has argued for relabeling the school subject of "religious education" as "worldview education."

My focus here is not on debates about the definition of "religion" nor on the question of how we should label what we are doing when we teach about religions and beliefs. Rather, my focus is on the worldview promoted by mainstream economic theory and embodied in the institutions of global capitalism and the lifestyle of consumerism; on the relationship of these components of "economic religion" to our current crisis of sustainability; and on the task of envisioning alternatives.

12.4 A Clash of Worldviews

During the 1980s former British Prime Minister, Margaret Thatcher, famously declared that "there is no alternative" to a deregulated and increasingly globalized market-based capitalism focused on the quest for economic growth. This view continues to be promoted by business executives, political leaders, and a media environment saturated with idealized images of a consumer lifestyle in a free-market world (Steger, 2009, pp. 102–103).

But is it true that there is no alternative?

The most common ways of measuring success and well-being in our world are quantitative in nature. Calculations of economic development routinely rely on such measures as the Gross Domestic Product, yet the GDP is merely a sum of national spending. It makes no distinctions between transactions that add to human and environmental well-being and those that diminish it. For example, the GDP counts pollution as a double gain. While pollution clearly detracts from the quality of life, the industrial activity that caused it in the first place and the money spent to clean it up both contribute to economic 'growth' as calculated by the GDP (Anielski, 1999). When standard economic indicators focus simply on the sheer amount of economic activity without making any qualitative distinctions between kinds of activity, we are presented with a distorted and deceptive picture of reality.

An alternative way of measuring economic outcomes and societal wealth is provided by the Genuine Progress Indicator or GPI. The GPI starts with the same data as the GDP, but it then incorporates environmental and social factors that are not measured by the GDP. For example, it adjusts for factors such as income inequality, adds factors such as the value of household and volunteer work, and subtracts factors such as the costs of crime and pollution. By including qualitative factors in its measurements, the GPI provides a more truthful picture of economic and social progress. While the GDP has continued to go up over the past several decades, the GPI shows that growth in the quality of life has been more or less stagnant since the 1970s (Anielski, 1999; Hayes, 2021).

This shift from a quantitative to a qualitative measure of well-being is at the core of a new more ecological approach to economic theory pioneered by Herman Daly and others (Daly, 1991, 2013; Daly & Cobb Jr., 1994; see also Hickel, 2021; Jackson, 2017, 2021; Kallis et al., 2020; Raworth, 2017). In contrast to mainstream growth-based economic theory, which largely ignores the biophysical realities of the planet, ecological economics understands the economy as a subsystem of the ecosphere. It argues that there is a limit to the total material production that the ecosystem can support and envisions a steady-state economy embedded within natural constraints (Rees, 2015, p. 1).

A steady-state economy entails stabilized population and per capita consumption. It can achieve higher levels of production and consumption through minimizing waste, but it cannot exceed ecological limits. The economy's "steady state" is indicated by stabilized (or mildly fluctuating) GDP. (While GDP is not a good indicator of qualitative well-being, it is a useful indicator of overall economic activity and environmental impact) (CASSE, 2021).

This steady-state is not to be confused with a stagnant state. A steady-state economy can be dynamic and continue to develop, constantly changing with the rise of new and the decline of "sunset" industries. But it aims for qualitative improvement instead of the perpetual quantitative growth that is overwhelming the carrying capacity of the natural environment (Rees, 2015, p. 8). The "success" of a steady-state economy is measured by its ability to maximize qualitative well-being rather than maximizing profits and GDP (Pizzigati, 2022).

While GDP growth remains vital for poor countries struggling to meet the basic needs of their populations, in wealthy countries the social and ecological costs of increasing growth have begun to outweigh the benefits. On this view, those of us who already participate in the prosperity of the developed world need to reject the goal of always having *more* and revolve our lives instead around having *enough*. In both wealthy *and* poor countries, more needs to be done to address vast and growing disparities of income and wealth (Marchese, 2022; Pizzigati, 2022). Among other things, this will involve more sharing of resources, a virtue that is promoted, at least in principle, by all of the world's great religious and ethical traditions. Yet discussions of sharing and redistribution are often derided today as appeals to "class warfare." Daly dismisses such rhetoric and argues that "real class warfare will not result from sharing, but from the greed of the elites who promote growth because they capture nearly all of the benefits from it, while 'sharing' only the costs" (cited in Pizzigati, 2022).

Proponents of ecological economics have set forth a number of concrete policy proposals for transitioning to a steady-state economy and thereby addressing two of the most pressing and interrelated challenges of our time: environmental collapse and economic inequality. These include various combinations of tax and banking reforms; implementation of cap-auction-trade systems for basic resources; proposals for minimum and maximum distributions of income to address poverty and limit the range of inequality while maintaining incentives and fair compensation for work; reforms of the World Trade Organization, World Bank, and International Monetary Fund; and re-regulation of international commerce (Daly, 2013; see also Daly & Cobb Jr., 1994; Boggs, 2012; Brown, 2009; Feffer, 2023; Paoletta, 2022; Rees, 2011, 2014).

Advocates of ecological economics recognize that they are calling for a profound shift in worldview – "the conceptual change in vision from the norm of a growth economy to that of a steady-state economy is radical" (Daly, 2013, p. 6). A key challenge facing any effort to create a sustainable future will be to imagine alternative conceptions of an abundant life and of the relationship between material wealth, consumption, and well-being (Grelle, 2018).

12.5 The World's Religions as Resources for Our Ecological and Moral Imaginations

One of the most necessary and practical responses to the crisis of sustainability precipitated by the growth-based global economic system is to articulate alternatives. It is at this point that teaching and learning about the world's traditional religions can play a valuable role. As ecofeminist theologian Sallie McFague (2001) has noted,

> Religions are in the business of envisioning counter-cultural alternatives on how to live rightly … generally these views do not propose materialism, consumption, and greed as virtues. On the contrary, altruism, sharing, compassion, self-limitation, detachment, and sacrifice are frequent motifs in most religions (p. 35).

The value of alternative visions is twofold. They provide a perspective from which to critique what is, and they provide a goal for which to strive; "they cut into our complacency, our selfishness, and our despair and make us realize that we are capable of – and want – something better" (p. 36).

A notable example of a religiously grounded moral critique of the present world order can be seen in *Towards a Global Ethic: An Initial Declaration of the Parliament of the World's Religions*, originally promulgated in 1993 and updated in 2018 in order to more explicitly address the issues of sustainability and care for the environment (Parliament, 2018; Renaud & Schweiker, 2021). The declaration argues that in the aftermath of the twentieth century's world wars, the collapse of fascism and Nazism, the breakdown of communism and colonialism, and the technological, economic, political, and cultural transformations wrought by globalization, "humanity has entered a new phase of history" (Parliament, 2018, p. 4). Against a background of "crisis in global economy, global ecology, and global politics," the declaration proclaims that there can be "no new global order without a new global ethic … Without such a fundamental consensus on an ethic, sooner or later every community will be threatened by chaos or dictatorship, and individuals will despair" (Parliament, 2018, pp. 3–5).

The declaration asserts that "a minimal *fundamental consensus* concerning binding *values*, *irrevocable standards*, and *fundamental moral attitudes*" – already exists in the teachings of the world's religions. While acknowledging that "this ethic provides no direct solution for all the immense problems of the world," it nonetheless does supply the basis for "a vision which can lead women and men away from despair, and society away from chaos" (Parliament, 2018, p. 3, emphasis in original). This vision begins with the fundamental demand that "every human being must be treated humanely," a demand long recognized by the world's great religious and ethical traditions in the form of the "golden rule" (Parliament, 2018, pp. 5–6). The declaration goes on to identify five "irrevocable directives" that are "implied" by this fundamental principle:

1. Commitment to a culture of non-violence and respect for life.
2. Commitment to a culture of solidarity and a just economic order.
3. Commitment to a culture of tolerance and a life of truthfulness.
4. Commitment to a culture of equal rights and partnership between men and women
5. Commitment to a culture of sustainability and care for the Earth. (Parliament, 2018, pp. 6–13).

This fifth directive calls attention to the over-exploitation and unjust use of finite natural resources. It claims that consumption must take into account the limits of what ecosystems can bear, and it asserts that "wherever development surpasses sustainable limits, we have the duty to speak up, to change our practices, and to moderate our lifestyles" (Parliament, 2018, p. 12). Of special note is its statement that

> Young people should be encouraged to appreciate that a good life is not a life of outsized consumption or amassing material possessions. A good life strikes a balance between one's needs, the needs of others, and the health of the planet. Education about the environment and sustainable living should become part of the school curricula in every country in the world (p. 12).[1]

The declaration concludes with a pledge to "work towards a transformation of consciousness," claiming that "Earth cannot be changed for the better unless we achieve a transformation in the consciousness of individuals and in public life" (p. 13).

The Global Ethic sets forth a compelling alternative to the materialistic and self-interested visions and values that dominate contemporary economics, politics, popular culture, and even increasingly education. While not ignoring the crimes committed in the names of all the world's religions or the role that religions have undeniably played in the creation or exacerbation of many contemporary global problems, the declaration helps make the case that religions have always been and must continue to be involved in efforts to ameliorate these problems. It acknowledges the complex and ambiguous role of religions in history and society, both as sources of conflict and as spiritual and moral resources in the pursuit of reconciliation, peace, justice, and now, sustainability.

Framing discussions of the present world order in terms of ethics has been a hallmark of both ecological economics and religious participants in the movement for global peace, justice, and environmental responsibility. Issues of economics, politics, and sustainability are set within the context of an overarching moral question: what is the present world order doing to and for people and the natural environment?

12.6 Rethinking Sustainability

A change in our present "system" – our prevailing economic worldview, our lifestyle of consumerism, and the accompanying practices and institutions – will require a *conversion in the way that we* think *about sustainability*. We can no longer assume that sustainability is about finding a way to preserve our present "system." Such use of the concept serves no purpose other than "greenwashing" – a public relations strategy that simply seeks to mask business as usual. If the concept of sustainability is to be retained it must be understood and implemented as a tool for criticizing and

[1] The Parliament of the World's Religions has also formed a Climate Action Task Force emphasizing the importance of global interfaith harmony in encouraging and supporting climate action by individuals and communities, governments at all levels, religious congregations, and for-profit and not-for-profit entities to reduce and eliminate greenhouse gas emissions by reforming patterns of production and consumption, decreasing reliance on fossil fuels, promoting renewable energy and energy efficiency, and promoting the sustainability of natural systems. https://parliamentofreligions.org/climate-action/

rejecting as *unsustainable* the growth-based economic status quo and as a spur to the development of more ecologically oriented alternatives.

This call for conversion alerts us yet again to the religious dimensions of current debates about sustainability. No less than mainstream economics, ecological economics – and environmentalism more generally – are grounded in comprehensive worldviews and hierarchies of value that provide people with a sense of meaning and purpose. Indeed, Robert H. Nelson has written of the "New Holy Wars" that that pit "economic religion" and "environmental religion" against one another in a struggle for the soul of contemporary societies (2010a).

Arguing about what we mean when we talk about sustainability is one of the main battlegrounds on which these wars are being fought. Mainstream economic theorists continue to argue that the solution to our problems is more growth. As once declared by Lawrence Summers, former chief economist of the World Bank and official in both the Clinton and Obama administrations, "There are no limits on the carrying capacity of the Earth that are likely to bind any time in the foreseeable future … The idea that we should put limits on growth because of some natural limit, is a profound error and one that, were it ever to prove influential, would have staggering social costs" (cited in Jackson & Jensen, 2022, p. 76). Yet, as Jackson and Jensen observe, the staggering social costs are already with us, not because we put limits on economic growth but because we failed to do so (p. 76).

The sort of "techno-optimism" (Jackson & Jensen, 2022, p. 21) that imagines we will invent our way to a sustainable future without dramatic realignment slides easily into "technological fundamentalism," described by Wes Jackson as "a religious-style faith in the ability of societies to solve problems with high energy and high technology, including the problems created by past use of that energy and technology... [T]his faith in technology allows people to avoid a simple reality: an economics of endless growth on a finite planet will end badly" (pp. 62–63).

What will it take to displace growth as the be-all and end-all of mainstream economic and political thinking? William Rees (2015) holds out the possibility that "some combination of good science, popular fear, and civic unrest will finally make of ecological economics an idea whose time has come" (p. 8). Yet Herman Daly (2013) himself frequently acknowledged that currently his proposals for transitioning to a steady-state economy "are beyond the pale politically" (p. 5). He surmised that "Only after a significant crash, a painful empirical demonstration of the failure of the growth economy" would his proposals or anything like them stand a chance of being enacted (pp. 5–6). In an interview a few months before he died in October, 2022, Daly was asked whether he was disappointed that his ideas had not been more successful in shifting the terms of the discussion. He said that what was disappointing was that much of the criticism directed at him did not claim that his presuppositions or reasoning were wrong, but that his ideas were "unrealistic" (Marchese, 2022).

But what is more unrealistic? To assume that we can carry on with business as usual and continue on the trajectory toward overshoot and collapse? Or to work toward the creation of a post-growth way of viewing and being in the world? What are the prospects for the emergence of a more qualitative orientation to economic decision making?

A 2021 study titled "Young People's Voices on Climate Anxiety, Government Betrayal and Moral Injury: A Global Phenomenon," surveyed 10,000 people, ages 16–25, across 10 different countries: India, Nigeria, Philippines, Portugal, Australia, US, UK, Finland, France, and Brazil. Nearly half said that their thoughts and feelings about the climate crisis have significantly impaired their ability to function: to eat, work, sleep, spend time in nature, have fun, and be fully present in their relationships (Marks et al., 2021; Wray, 2022a).

This experience of "eco-anxiety" has begun to impact young people's visions for their lives – their conceptions of success; their decisions about whether or not to have children; their perceptions of the meaning and purpose of education and work; and their professional and career ambitions (Wray, 2022a, b). A 2022 poll shows that 2 out of 3 members of Gen Z are anxious about climate issues, and almost as many feel it's important for their employers, and future employers, to show true environmental leadership. Slightly over half would take a pay cut if it meant working for an employer that better aligns with their ethics (BUPA, 2022; Wray, 2022a).

Rather than continue with an unquestioning allegiance to the ethic of consumerism – more money, more status, more stuff – many are exploring more holistic ways of thinking about what it means to grow. These shifts in ambition and disengagements from traditional career expectations are freeing people from a focus on themselves alone and leading them to recognize the interdependence that exists between their own well-being and that of the human and natural communities of which they are a part (Wray, 2022a). Clearly these sorts of surveys are not predictive of the future, but they indicate a willingness on the part of some young people to rethink several of the basic assumptions of our dominant system.

Wes Jackson and Robert Jensen (2022) are less upbeat about current trends, noting that even if today's various progressive social movements were to be as successful as possible, the human future is still likely to be "gritty and grim" (p. 11). They argue that in the present circumstances, the anxiety and despair that so many people of all ages are experiencing is not irrational. It should not be minimized nor waved away with platitudes. Even so, they counsel that "Analysis (no matter how grim) and action (no matter how slim the chances of success) are antidotes to despair" (pp. 4–5). The challenge here is to steer a course between nihilism on the one hand – the idea that nothing can be done – and wishful thinking on the other hand – the idea that we can do whatever we set our minds to. "The task is to recognize what is impossible and what is – or at least might be – possible, not in the abstraction of theory, but in the concreteness of the world" (p. 4).

12.7 Contributions of Religious Literacy Education

As the twenty-first century continues to unfold, a range of religious and secular worldviews are vying for influence in the public sphere. Helping students to grasp the history, content, and consequences of competing religions and beliefs is essential for preparing them to address the multiple crises that our world is facing.

An academic approach to education about religions and beliefs aims "to introduce students to the vast array of faith-based expressions that exist within and between traditions with the aim of deepening understanding about religious diversity and the roles that religion plays in political, economic, and cultural life across time" (AAR, 2010, p. 4). At the heart of "religious literacy" is "the ability to discern and analyze the intersections of religions with social, political, and cultural life" (AAR, 2010, p. 4).

There is widespread agreement that schools should teach about diverse religious and secular perspectives in an unbiased fashion; they must avoid proselytizing (Bird et al., 2016, pp. 306–329). And yet it might be argued that schools are guilty of proselytizing on behalf of "economic religion" when their required courses on economics focus on the neoclassical model of economic behavior to the exclusion of alternative ecological and religious perspectives on economic life (Nord, 2010, pp. 47–52, 228–231; Nord & Haynes, 1998, pp. 105–114; Moore, 2007, pp. 171–173). Likewise, it might be argued from the other side that schools sometimes promote the agenda of "environmental religion" in various courses and student activities outside the economics classroom (Nelson, 2010a, p. 1). In this as in so many other areas, schools find themselves caught up in the culture wars that have become common in pluralistic democratic societies. Neither form of proselytization is appropriate. An *academic* as opposed to a *devotional* approach should encourage student awareness by informing them about multiple religious and secular worldviews without seeking to impose, promote, or denigrate any particular viewpoint (AAR, 2010, pp. 7–8).

But as I have tried to show, helping students to understand the disparate assumptions and values that are brought to bear on discussions of sustainability is not so much a matter of proselytizing as it is a matter of basic education. Without this sort of religious literacy, students will be unable to understand the pressing environmental, economic, and social-ethical issues that are at stake in current debates about sustainability.

Education about alternative religious and secular worldviews may also help to nurture hope in the face of growing levels of eco-anxiety and despair that many are experiencing. Frederick Bird (2022) has argued that hope is best understood as the disposition to realistically anticipate possibilities so as to address present challenges when the future is uncertain and unknown (pp. 3–11). On this view, hope is not a state of mind or emotion that depends on factors or circumstances outside ourselves, but rather a disposition or virtue that can be cultivated.

Hope is not the same as optimism. Optimism involves a confident assumption that there is a high probability of positive outcomes in the future and that everything will turn out for the best. By contrast,

> hope enables us to manage crises in the present when the future is unknown and uncertain without being overcome either by despair or wishful thinking. Hope empowers people. It acts as a generative force, facilitating the flow of ideas and energy. It functions to open up our minds, hearts, and will. It serves as an ongoing invitation to explore, to think and re-think, and to try again (Bird, 2022, p. 9).

This disposition to hope – to realistically anticipate possibilities – helps to "initiate changes that open expectations for others" (p. 7).

The possibility of conceiving economic life in any terms other than those of the neoclassical growth paradigm is something that has never occurred to a surprisingly large number of people. The idea that it might be possible to conceive of an abundant life in other than materialistic terms is a novel idea for many. Perhaps teaching and learning about alternative worldviews can contribute not only to the imagination, but also to the embodiment, of a new and genuinely sustainable manner of living in the world.

References

AAR Religion in the Schools Task Force, Diane L. Moore, Chair. (2010) Guidelines for teaching about religion in k-12 public schools in the United States. *American Academy of Religion*. https://aarweb.org/common/Uploaded%20files/Publications%20and%20News/Guides%20and%20Best%20Practices/AARK-12CurriculumGuidelinesPDF.pdf

Anielski, M. (1999). *The genuine progress indicator – A principled approach to economics*. Retrieved March 29, 2023, from https://www.pembina.org/reports/gpi_economics.pdf

Bird, F. (2022). *The generative power of hope: Anticipating possibilities in times of crises*. Springer.

Bird, F., Twiss, S. B., Pedersen, K. P., Miller, C. A., & Grelle, B. (2016). *The practices of global ethics: Historical backgrounds, current issues, and future prospects*. Edinburgh University Press.

Boggs, C. (2012). *Ecology and revolution: Global crisis and the political challenge*. Palgrave Macmillan.

Brown, L. R. (2009). *Plan b 4.0: Mobilizing to save civilization*. W.W. Norton & Company.

BUPA/British United Provident Association. (2022, January 13). *Gen z seek ethical workplaces as environ-mental health burden bites*. Retrieved March 29, 2023, from https://www.bupa.com/news/press-releases/2022/gen-z-seek-ethical-workplaces-as-environ-mental-health-burden-bites

CASSE/Center for the Advancement of the Steady State Economy. (2021). *Definition of steady state economy*. Retrieved March 30, 2023, from https://steadystate.org/discover/definition-of-steady-state-economy/

Council for Economic Education. (2010). *Voluntary national content standards in economics*, 2nd edn. Retrieved March 31, 2023, from https://www.councilforeconed.org/wp-content/uploads/2012/03/voluntary-national-content-standards-2010.pdf

Cox, H. (1999, March). The market as god: Living in the new dispensation. *The Atlantic*. http://www.theatlantic.com/magazine/archive/1999/03/the-market-as-god/306397/

Cox, H. (2016). *The market as god*. Harvard University Press.

Daly, H. E. (1991). *Steady-state economics* (2nd ed.). Island Press.

Daly, H. E. (2013, October 28). *Top 10 policies for a steady-state economy*. Retrieved March 30, 2023, from https://steadystate.org/top-10-policies-for-a-steady-state-economy/#:~:text=Top%2010%20Policies%20for%20a%20Steady%20State%20Economy,free%20capital%20mobility%2C%20and%20globalization.%20...%20More%20items

Daly, H. E., & Cobb, J. B., Jr. (1994). *For the common good: Redirecting the economy toward community, the environment, and a sustainable future* (2nd ed.). Beacon Press.

Feffer, J. (2023, January 18). From the unsustainable here to the sustainable there: Economic growth is killing the planet. How do we engineer an alternative? *Foreign Policy in Focus*. Retrieved March 29, 2023, from https://fpif.org/from-the-unsustainable-here-to-the-sustainable there/?emci=fad7b1a6-3c98-ed11-994c-00224832eb73&emdi=daf03ead-d698-ed11-994c-00224832eb73&ceid=3990497

Foltz, R. (2007). The religion of the market: Reflections on a decade of discussion. *Worldviews, 11*, 135–154.
Global Footprint Network. (2023). *Earth overshoot day*. Retrieved March 30, 2023, from https://www.overshootday.org/
Grelle, B. (2018). Worldviews, ethics, and ecology: Sustainability as a context for religious education. In J. Astley, L. J. Francis, & D. W. Lankshear (Eds.), *Values, human rights and religious education* (pp. 189–206). Peter Lang.
Hayes, A. (2021, July 25). *Genuine progress indicator (GPI): Definition, formula, vs. GDP*. Retrieved March 30, 2023, from https://www.investopedia.com/terms/g/gpi.asp
Heinberg, R. (2012, March 15). *What is sustainability?* Retrieved March 30, 2023, from https://www.postcarbon.org/publications/what-is-sustainability/
Hickel, J. (2021). *Less is more: How degrowth will save the world*. Windmill Books.
Jackson, R. (2014). *Signposts – Policy and practice for teaching about religions and non-religious worldviews in intercultural education*. Council of Europe.
Jackson, T. (2017). *Prosperity without growth: Foundations for the economy of tomorrow* (2nd ed.). Routledge.
Jackson, T. (2021). *Post growth: Life after capitalism*. Polity Press.
Jackson, W., & Jensen, R. (2022). *An inconvenient apocalypse: Environmental collapse, climate crisis, and the fate of humanity*. University of Notre Dame Press.
Kallis, G., Paulson, S., & D'Alisa, G. (2020). *The case for degrowth*. Polity Press.
Loy, D. (1997). The religion of the market. *Journal of the American Academy of Religion, 65*(2), 275–290.
Marchese, D. (2022, July 17). This pioneering economist says our obsession with growth must end. *New York Times Magazine*. Retrieved March 31, 2023, from https://www.nytimes.com/interactive/2022/07/18/magazine/herman-daly-interview.html
Marks, E., Hickman, C., Pihkala, P., Clayton, S., Lewandowski, E. R., Mayall, L. E., Wray, B., Mellor, C., & van Susteren, L. (2021, September 7). *Young people's voices on climate anxiety, government betrayal and moral injury: A global phenomenon*. Retrieved March 31, 2023, from https://papers.ssrn.com/sol3/papers.cfm?abstract_id=3918955
McFague, S. (2001). *Life abundant: Rethinking theology and economy for a planet in peril*. Fortress Press.
Mohr, C. (2017, March 1). Review of The market as god, by Harvey Cox. *Friends Journal*. Retrieved March 31, 2023, from https://www.friendsjournal.org/the-market-as-god/
Moore, D. L. (2007). *Overcoming religious illiteracy: A cultural studies approach to the study of religion in secondary education*. Palgrave Macmillan.
Nelson, R. H. (2001). *Economics as religion: From Samuelson to Chicago and beyond*. The Pennsylvania State University Press.
Nelson, R. H. (2010a). *The new holy wars: Economic religion vs. environmental religion in contemporary America*. The Pennsylvania State University Press.
Nelson, R. H. (2010b, July 20). Religion, economics, and the market paradox. *Religion & Liberty, 12*(1). Retrieved March 31, from https://www.acton.org/pub/religion-liberty/volume-12-number-1/religion-economics-and-market-paradox
Nord, W. A. (2010). *Does god make a difference? Taking religion seriously in our schools and universities*. Oxford University Press.
Nord, W. A., & Haynes, C. C. (1998). *Taking religion seriously across the curriculum*. Association for Supervision and Curriculum Development.
OSCE/Organization for Security and Co-operation in Europe. (2007). *Toledo guiding principles on teaching about religions and beliefs in public schools*. ODIHR.
Paoletta, K. (2022, April 18/25). The end of growth. *The Nation*, pp. 20–23, 31.
Parliament of the World's Religions. (2018). *Towards a global ethic: An initial declaration of the parliament of the world's religions*. Retrieved July 31, 2023, from https://parliamentofreligions.org/wp-content/uploads/2023/05/Global-Ethic-PDF-2020-Update.pdf

Pizzigati, S. (2022, November 14). *An economist that future economists – and societies – will dare not ignore*. Retrieved April 7, 2023, from https://inequality.org/great-divide/a-economist-that-future-economists-and-societies-will-dare-not-ignore/

Raworth, K. (2017). *Doughnut economics: Seven ways to think like a 21st century economist*. Chelsea Green Publishing.

Rees, W. (2011, April). *Toward a sustainable world economy*. Retrieved April 2, 2023, from https://www.ineteconomics.org/research/research-papers/toward-a-sustainable-world-economy

Rees, W. (2014, June 12). *Avoiding collapse: An agenda for sustainable degrowth and relocalizing the economy*. Retrieved April 2, 2023, from https://www.policyalternatives.ca/publications/reports/avoiding-collapse

Rees, W. (2015, June). *Economics vs. the economy*. Retrieved April 2, 2023, from https://great-transition.org/images/Rees-Economics-vs-Economy_1.pdf

Renaud, M., & Schweiker, W. (2021). *Multi-religious perspectives on a global ethic: In search of a common morality*. Routledge.

Robinson, W. I. (2014). *Global capitalism and the crisis of humanity*. Cambridge University Press.

Smart, N. (1995). *Worldviews: Crosscultural explorations of human beliefs* (2nd ed.). Prentice Hall.

Steger, M. B. (2009). *Globalization: A very short introduction*. Oxford University Press.

Valk, J. (2021). *Worldviews: A comprehensive approach to knowing self and others*. Palgrave Macmillan.

Wackernagel, M., & Rees, W. (1996). *Our ecological footprint: Reducing human impact on the earth*. New Society Publishers.

Wray, B. (2022a, March). *Climate distress is reshaping young people's professional ambitions*. Retrieved, March 28, 2023, from https://gendread.substack.com/p/climate-distress-is-reshaping-young?utm_source=substack&utm_medium=email

Wray, B. (2022b). *Generation dread: Finding purpose in an age of climate crisis*. Alfred A. Knopf Canada.

Dr. Bruce Grelle is Professor Emeritus in the Department of Comparative Religion and Humanities at California State University, Chico. He served on the American Academy of Religion's Task Force on "Religion in the Schools" and was co-founder of the AAR's program unit on "Religion and Public Schools: International Perspectives."

Chapter 13
Concluding Remarks and Future Directions

Olof Franck and Sally Windsor

13.1 Conclusion

In gathering contributions for this book, we anticipated that it would contribute to global conversations and debates on how sustainability issues could more deeply, and explicitly, be integrated into varied forms of religious education. The concepts inherent in the title of this book, *religion*, e*ducation,* and s*ustainability* are deep and wide and the possibilities for research and development in various directions seems to be vast. We recognise that in a book such as this we cannot include all the religious perspectives on education and sustainability; however, we have actively sought a diversity of contributions. In the foregoing chapters one can see the chapters represent a dynamic area that offer glimpses of a disciplinary arena that seems difficult to fully grasp. Nevertheless, we present a range of issues related to how education can teach about sustainability from different religious, cultural, and educational perspectives and with reference to various research approaches. The authors in this book show commitment, wisdom and creativity relating to environmental, social and economic sustainability.

As the world is changing so rapidly on both global and national levels, and serious challenges are looming, the chapters in this book provide a grounding in areas that should be further researched. We believe there are three obvious paths to take for continuous forward-looking research areas at the intersection of religion, education, and sustainability. First, it is crucial for the future that notions of what a good life and a sustainable society can, and should, be in historical and contemporary religious traditions is further explored. It is important (and interesting) to study how perspectives on social, economic, and environmental sustainability are expressed

O. Franck (✉) · S. Windsor
Department of Pedagogical, Curricular and Professional Studies, Faculty of Education, The University of Gothenburg, Gothenburg, Sweden
e-mail: olof.franck@gu.se

S. Windsor, O. Franck (eds.), *Intersections of Religion, Education, and a Sustainable World*, https://doi.org/10.1007/978-3-031-81809-7_13

and can be traced in both older and newer religious texts and teachings. We must analyse how religious or theological language gives voice to a quest for care, for creation, for all living beings life together. Such an analysis needs to be critical as well as constructive, paying attention to claims to truth and knowledge as expressed in religious and theological contexts.

Second, a critical and constructive analysis should also explore how ethics and morality are formulated, critiqued, and defended in different religious and theological arenas – in different times, in different societies, in different disciplinary discourses. What is the place and importance of environmental and social values internal, and external, to religious and theological contexts? How do discussions about democracy, justice and equality look in such arenas, and how do they relate to discussions in secular contexts?

Third, it is important to study how people's concerns about the future, about climate threats and about societal polarization in different areas, are noticed and interpreted within the framework of religious and theological traditions and movements. The existential question of whether and how it is possible to find hope and meaning in an ethically and epistemically troubled world is a burning one, especially for young people. It is a relevant question for both religious and secular worldviews to pay attention to, interpret and develop.

The chapters in this book can each be related to these three areas of study of religions and worldviews, where research and education on different dimensions of sustainability and sustainable development are discussed. In doing so, they make important contributions to knowledge about how religion, education and sustainability can be understood in the present time.

Olof Franck is Professor in Subject matter education, specializing in the Social Sciences and Associate Professor in Philosophy of Religion at the Department of Pedagogical, Curricular and Professional Studies, University of Gothenburg. His research and teaching interests mainly belong to the fields of religion education, ethics education, ethics, philosophy of religion and religious studies, as well as sustainability education with a focus on social sustainability, ethics and values.

Sally Windsor is an Associate Professor in pedagogical work with a focus on sustainability and international education. Her research and teaching focuses upon teacher training and professional development in terms of interdisciplinary and cross-curricular teaching for social sustainability and global citizenship education.

MIX
Papier aus verantwortungsvollen Quellen
Paper from responsible sources
FSC® C105338
FSC
www.fsc.org

If you have any concerns about our products,
you can contact us on
ProductSafety@springernature.com

In case Publisher is established outside the EU,
the EU authorized representative is:
Springer Nature Customer Service Center GmbH
Europaplatz 3, 69115 Heidelberg, Germany

Printed by Libri Plureos GmbH
in Hamburg, Germany